Introduction

Getting Started

Housing & Protection

Nutrition & Feeding

Health & Wellness

Breeding Gouldian Finches

Gouldian Finches
Care, Breeding & Genetics

Genetic Considerations

Common Challenges & Solutions

Questions & Answers

Conclusion

Glossary of Terms

Forms Library

Gallery of Colours & Mutations

Appendices

Index of Key Words

Gouldian Finches - Care, Breeding & Genetics
Tony Hanks

All rights reserved
Copyright © 2025 by Anthony Hanks, OAM

No part of this publication may be reproduced, distributed, or transmitted in any form or by any means, including photocopying, recording, or other electronic or mechanical methods, without the prior written permission of the author, except in the case of brief quotations embodied in critical reviews and certain other non-commercial uses permitted by copyright law.

Published by Gouldian Care

www.gouldiancare.com
ISBN: 978-1-7644756-0-0

Graphics:	All photos, diagrams, tables and forms are by Tony Hanks
Websites:	https://www.gouldiancare.com | https://books.by/tony-hanks
BISAC Codes:	PET002000 | SCI029000 | TEC003020
Keywords:	Gouldian | Gouldians | Finches | Gouldian Breeding | Gouldian Genetics | Gouldian Mutations | Aviculture |
Editions:	First edition - 2025
File Ref:	1.10

Cover Photo:

Front to back . .
- Red Headed Yellow Back
- Yellow Headed Wild-Type
- Black Headed Wild-Type
- Black Headed White Breasted
- Black Headed Dilute (SF Yellow Back)

Disclaimer: The opinions expressed in this book are solely those of the author and do not necessarily reflect the views or positions of any organization or individual associated with the author. Readers are advised to approach the content critically and independently, as it represents the author's personal perspective and interpretation. Any reliance on the information provided herein is at the reader's own discretion.

CONTENTS

1: Introduction 15
    The Allure of Gouldian finches: A Brief History 15
        Head Colour Proportions in Nature
    Why choose Gouldian finches? 16
    Understanding Their Natural Habitat and Behaviour 16
    Breeding Period & Life Expectancy 17
        Breeding Maturity
        Native Habitat Temperature Range
        Hen Fertility
        Life Expectancy
    Average Body Size 18
    Basic Genetic Concepts 18
    Terminology 19
    Behaviour Related to Head Colours 19
        Red-Headed Behaviour
        Black-Headed Behaviour
        Yellow-Headed Behaviour

2: Getting Started 21
    Identifying Healthy Birds 21
        Key Indicators to Look For
    Sexing Gouldian Finches 22
        Natural Wild-Types
        Mutations
    Quarantining New Birds 23
        Benefits
        Period of Quarantine
    Gouldian Finches in Mixed Collections 23
        Recommended Species
        Aggressive Species to Exclude
    Essential Supplies and Equipment 24
        Perches
        Nesting Boxes
        Food Dispensers

> *Water Dispensers*
> *Bathing Facilities*
> *Cuttlefish holder*
> *Shell Grit Container*
> *Nesting Material*
> *Catching Net*
> *Transport Cage*
> *Hospital Cage*
> *Stress Perches*

Climate & Temperature 28
> *Minimum Temperatures*
> *Maximum Temperatures*

Leg Rings and Record Keeping for Birds 28
> *Ring Size*
> *Split or Fixed*
> *Accurate Records*

3: Housing & Protection 31

Choosing an Aviary, Flight, Cage, or Cabinet 31
> *The Housing Options*
> *Factors to Consider*
> *Flight or Cabinet Sizes*

Guideline for Number of Birds 32
> *Volume or Area*
> *Housing Guideline*

Wind Protection 33
> *Danger of Cold Draughts*
> *Wind Barriers*

Location & Aspect 34
> *Direction*
> *Natural Sunlight*
> *Cold Weather Protection*
> *Hot Weather Protection*

Sunlight & UV Light 35
> *Vitamin D Synthesis*
> *Regulating Circadian Rhythms*
> *Mental Stimulation and Well-being*
> *Full-Spectrum Lighting in a Flight or Cabinet*
> *Recommendation*

Cleaning and Maintenance 36
Floors of Aviaries & Flights 36
> *Concrete Floors*
> *Natural Earth Floors*
> *Sand Floors*
> *Wooden Floors*
> *Wire Mesh Floors*

    Gravel Floors
    Recommendation
  Floors of Breeding Cabinets 37
    Sand Filling
    Paper Lining
    Pine Shavings
    Animal Litter
    Wire Mesh Floors
    Recommendation
  Custom Design for a Flight 39
    Planning and Design
    Building Materials
    Construction
    Interior Setup
    Example of a Flight Design
  Custom Design for a Breeding Cabinet 41
    Space Requirements
    Materials and Construction
    Ventilation and Lighting
    Accessibility and Maintenance
    Water Options
    Feed Options
    Rotating Coop Cups
    Floor Trays
    Foraging Trays
    Nesting Areas
    Nesting Material
    A Dry Floor
    Catching Dividers
    Temperature Control
    Supplementary Lighting
    The Benefits of Custom Designed Breeding Cabinets
    Conclusion
  Positioning of Perches 48
    Bird Behaviour and Roosting
    Minimum Measurements
  Quick Reference Table - Useful Standard Dimensions 50

4: Nutrition & Feeding 51

  Basic Diet 51
    Mixed Seed
    Austerity Seed
    Greens & Grains
    Seed Storage
  Supplements 53
    Calcium Supplements

Cuttlebone
Eggshells
Multivitamins
Vitamin D3
Amino Acids
Combined Amino Acids and Vitamins
Shell Grit
Electrolytes
Omega 3 & 6
Iodine
Apple Cider Vinegar (ACV)

Recommended Levels for Vitamin Supplements 57
   Underdosing & Overdosing
   Recommended Concentrations
   Recommended Level for Iodine
   Other Sources of Vitamins & Minerals
   Final Delivery Concentrations
   Undisclosed Concentrations

Feeding Tips 60
   Live Food
   Dried Live Food
   Vegetables
   Seeding Grasses
   Foraging Trays
   Boiled Eggs
   Egg & Biscuit Mix
   Blended Calcium Grit Products

Soaked & Sprouted Seed 65
   Soaked Seed
   Recommended Procedure for Soaked Seed
   Sprouted Seed
   Recommended Procedures for Sprouted Seed

Quick Reference Table - Annual Feeding Timetable 67

Special Diets for Special Birds 69
   Supplements for the Blue-Backed Mutation

Containers for Feeding Seed 69
   Open Dishes
   Feed Hoppers
   The Choice of Seed Hopper Design

Options for Fresh Water Supply 70
   Water Bowls
   Automatic Waterers
   Water Bottles
   Water Valves
   Water Fountains

Feeding Stations in an Aviary or Flight 71
   Components Included

Positioning
  Dimensions
  Nutrition Data 72
  Standard Care Plan 72

5: Health & Wellness 73

  Clean Environment 73
  Dry Environment 73
  Annual Cleaning 74
  Common Health Issues 74
   Respiratory Infections
   Bacterial Infections
   Viral Infections
   Fungal Infections
   Mite Infestations
   Feather Loss
   Coccidiosis
   Giardia
   Fractures & Trauma
   Eye Infections
   Nutritional Deficiencies
   Inadequate Natural Light
   Excessive Melanin
   Vitiligo
   Egg Binding
   Abnormal Head Movements
  Going Light 80
   Causes
   Symptoms
   Treatment
  Mutation Health Issues 81
  Flow Chart for Disease Treatment 82
  Medications 83
   Methods of Administration
   Dosage Rates
   Remote Dosing
   Off-Market
   International Products
   Safety Precautions
  Quick Reference Table - Summary of Diseases 85
  Use of a Hospital Cage 88
   Design
   Food & Water
   Advantages
   Treatment Notes

Preventive Care 89
   Routine Annual Medications
   Routine Trimming of Claws
Health Problems for Blue Gouldians 91
Apple Cider Vinegar 91
   Benefits of ACV
   Selecting the Right ACV
   Guidelines for Use of ACV
When to Consult a Veterinarian 92
Quick Reference Table - Measurements 94

6: Breeding Gouldian Finches 95

Importance of Genetics 95
Understanding the Breeding Cycle 95
   Annual Timetable
Colonies or Pairs 96
   Breeding Gouldians in Colonies
   Breeding Gouldians in Pairs
Creating the Optimal Breeding Environment 97
   Preparation for Breeding
   Suitable Nest Boxes
   Preparation of Nest Boxes
   Installation of Nest Boxes
Pairing and Mating: Selecting Compatible Pairs 99
   Close Relatives
   Ages
   Head Colours
Pair Bonding 100
   Formation of Pair Bonds
   Impact on Breeding Success
   Challenges to Pair Bonding
Predictable Progeny 101
   The Benefit of Predicting
   Breeding Plans Based on Predicted Progeny
   Example of a Non-recommended Pairing
   Examples of Breeding Plans for Mutations
Courtship 105
   Male Performance
   Female Readiness
Timetable for Hatching and Independence 106
   Egg-Laying
   Date Calculations
   Incubation
   Fledglings
   Independence

  Next Breeding Cycle
 The Nodules on Nestlings 109
  Purpose
  Differences Between Mutations
 Identifying Chicks in the Nest.. 109
 Using Foster Parents.. 110
  Higher Survival Rates
  Imprinting with Bengalese
  Gouldians as Foster Parents
 Candling Eggs to Check for Fertility 111
  Steps to Candle Finch Eggs
  Results When Candling Eggs
 Record Keeping for Breeding 112
  Computer or Paper Records
 Line Breeding... 114
  Goals of Line Breeding
  Risks of Line Breeding
  Managing Line Breeding
 Mating Preferences Related to Head Colour.. 115
  Assortative Mating
  Offspring Survival Rates & Gender Mixes
  Behavioural Choices
  Dominant Doesn't Dominate
 Breeding Results for Mutations.. 116
  Average Breeding Results
  Reasons for Reduced Breeding of Mutations
 Selling Offspring 117
  Readiness to Sell or Move
  Friends and Bird Club Members
  Bird Expos
  Bird Dealers or Local Pet Stores
  Online Marketplaces
  Information to Accompany a Sale

7: Genetic Considerations 119

 Why Understand Genetics.. 119
 Mendelian Inheritance.. 119
 Sex Chromosomes of Birds 120
 Genetics of Each Bird 120
  Computer or Paper Records
  Phenotype Summaries
  Genotype Summaries
  Single Factor or Double Factor
  Genotype Examples
 Breeding for Health.. 122

Colour Mutations 123
    Genotype Conventions
    Head Colours
    Breast Colours
    Back Colours
    Blue-Backed Mutation
    Yellow-Backed Mutations
    Yellow Back
    Australian Yellow
    Identifying the Yellow Backed Version
    Australian Recessive Dilute
    Cinnamon
    Seagreen
    Fallow
    Lutino
Single vs Double Factor Mutations 128
Combined Mutations 128
    White-Breasted Yellow Back
    White-Breasted Blue
Secondary Mutations 129
    Pastel Blue
    Silver
    White-Breasted Silver
    Australian Recessive Dilute Blue
    Ivory
    Satine
    Australian Variegated Blue
Head Colours in Mutations 132
    Changed Head Colour Appearances
Dominant & Recessive Traits 132
    Dominant Genes
    Recessive Genes
Quick Reference Table - Head Colours in Gouldian Mutations 133
Genetic Inheritance in Gouldians 134
Sex-Linked Inheritance 134
    Z & W Chromosomes
    Impact of Single & Double Factors
Quick Reference Table - Single & Double Factors in Mutations 135
Co-dominant Sex-Linked Inheritance 135
Autosomal Inheritance 135
    Breast Colour as an Example
Genetics of Mutations 137
    Phenotype Abbreviations
    Genotype Abbreviations
Tip of Beak Colours 137

Genetic Prediction Software 138
Quick Reference Table - Genetics of Gouldian Finches 139
Progeny Predictions 146
   Punnett Squares for 1 Trait
   How to Use Punnett Squares
   Examples of Calculated Outcomes
   Reference Tables for Progeny Predictions
Quick Reference Table - Gouldian Finch Progeny Predictions 150
The Science of Gouldian Colours 163
   Melanins
   Carotenoids
   Pteridines
   Genetic Influences on Colouration
   Environmental Influences on Colouration
   Pigment Changes in Gouldian Mutations

8: Common Challenges & Solutions 167

Recognizing and Addressing Breeding Issues 167
   Chick Tossing
   Excessive Heat
Health Problems and Treatments 168
   Air Sac Mites
   Injuries Due to Predators
   Vitamin D3 Deficiency
   Iodine Deficiency in Feather Loss
Behavioural Concerns and Solutions 172
   Territorial Disputes
   Abnormal Head Movements
   Poor Breeding Results
Housing Concerns and Solutions 174
   Webbing Moths in Seed Hoppers
   Diatomaceous Earth
   Building an Aviary: When Bigger is Not Always Better
Supplying Live Food 177
   Getting Started
   Establishing the Colony
   Breeding Cycle
   Collecting Larvae
   Feeding Methods

9: Questions & Answers 179

   Frequently Asked Questions

10: Conclusion 187
Recap of 12 Essential Care and Breeding Tips 187
Encouragement and Support for Finch Enthusiasts 189
Join a Finch Club
Join an Online Discussion Group
Join a Facebook Group
Resources for Further Learning

11: Glossary of Terms 191
Definitions & Meanings 191
Abbreviations Used 199
Ratios in Aviculture 201

12: Forms Library 203
Bird Log for Gouldian Finches 204
Annual Plan for Gouldian Finches 205
Decision Tree for Health Diagnosis 206
Dosage Rates 207
Health Treatment Notes 208
Breeding Date Calculations 209
Breeding Log for Gouldian Finches 210
Punnett Squares - 1 Trait - Predicted Progeny 211
Punnett Squares - 2 Traits - Predicted Progeny 212
Breeding Plan Based on Predicted Progeny 213
Offspring Log for Gouldian Finches 214
Sale Transfer Form 215
Reference Table - Gouldian Genetics Abbreviations 216
Bird Room Reference 217
Standard Care Plan for Gouldians 218

13: Gallery of Colours & Mutations 219
Photo Gallery 220

14: Appendices 225
Nutrition Facts 226
Vitamins & Minerals in Supplements 227
Popular Medications & Supplements 229
Recommended Reading 232
Recommended Resources 232

Products and Suppliers 232
Food Preference Trials 234
    Individual Seeds
    Sprouted Greens
References Used in This Book 236

15: Index 239

Fast Find Index 240
A Special Thank You 247

Tony Hanks, OAM

Tony Hanks lives on the east coast Australia and has experience with Gouldian finches over a period of 50 years.

As a teenager he was introduced to aviculture with Zebra finches and soon graduated to what were then seen as "more difficult" Gouldians.

Over the years Tony has constructed aviaries at five different locations as he moved house and each of these was designed for the specific needs of Gouldians. As he progressed from one to another his knowledge grew as he learned from the advice of colleagues and the experience gained.

He holds two professional degrees in science, including a doctorate. One area of interest from his academic background is the study of genetics and inheritance, which is very relevant to the breeding of Gouldians.

In 2010 Tony Hanks was recognized in the Queen's Birthday Honours List. He was awarded the Order of Australia Medal for "services to the community" by the Governor General of Australia.

Tony Hanks has had many articles and papers published internationally.

The Author

Gouldian Finches - Care, Breeding & Genetics

Introduction

Gouldian finches (Erythrura gouldiae), with their vibrant colours and engaging personalities, are captivating birds. This book offers a detailed guide on their care and breeding to help both hobbyists and professional breeders nurture these stunning birds.

The Allure of Gouldian finches: A Brief History

Gouldian finches, also known as Lady Gould finches or Rainbow finches, originate from the grasslands of Northern Australia. Named after his wife by the British ornithologist John Gould, they have been of interest to bird enthusiasts since their identification in the 19th century.

They are one of the 19 varieties of Australian grass finches and there are three forms of Gouldians with different head colours among the natural wild-type birds: Red, black and yellow headed. Some people refer to the last of these as "Orange-Headed" because the colour tone is more in that direction, but "Yellow-Headed" is the accepted naming convention.

The males (cocks) usually have more vivid colours than the females (hens), with heads that are red, black, or yellow; purple chests; yellow bellies; green backs; and blue tails. Females have more muted colours, but they are still extremely attractive visually.

Head Colour Proportions in Nature

The genes for black and yellow heads are recessive to red in different ways, which will be explained later. Despite this, red-headed Gouldians do not dominate in nature. The most common wild-type Gouldians have black heads. This is influenced by factors such as a preference for breeding with the same head colour, the increased aggressiveness of red-headed birds, their higher susceptibility to predators, and their shorter life expectancy by nearly two years.

Normal Gouldian head colours

In addition to the colour variations found in wild populations, aviculturists have also developed other colour mutations, which will be discussed in a later chapter of this book.

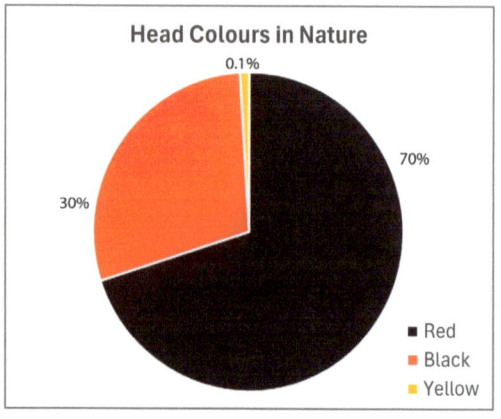

Why choose Gouldian finches?

Gouldian finches are well known for their striking colours and they are also social birds living in flocks that are interesting to observe.

The care requirements for Gouldian finches are specific, but not overly demanding. They require a diet rich in seeds and greens, along with access to fresh water at all times. Breeding Gouldian finches can be particularly rewarding because they are attentive parents and their chicks grow rapidly under proper care.

Understanding Their Natural Habitat and Behaviour

Gouldian finches thrive in warm climates, mirroring their native habitat, which consists of tropical savannas and woodlands. Despite their delicate size and appearance, these finches are resilient and adaptable if their basic needs are met.

The native habitat is the Kimberley region in Western Australia, stretching into part of the Northern Territory. In the major centres of Wyndham and Kunanurra the average temperature range is from a summer high of 39C in November to a winter low of 15C

Gouldian habitat in the Kimberley region of Western Australia and Northern Territory

Introduction

in July (102 to 59F). However, some individual days can be as high as 46C and as low as 5C (115 to 41F).

Understanding the social and vocal behaviours of Gouldian finches is essential for their proper care. Generally, these birds exhibit peaceful behaviour, but may become territorial during the breeding season.

Colours of Normal Gouldians

	Cock (Male)			Hen (Female)		
Head	Red	Black	Yellow	Red	Black	Yellow
Chin	Black			Black		
Nape	Blue			Blue Green		
Breast	Purple			Mauve		
Belly	Yellow			Light Yellow		
Back	Green			Green		
Tail	Blue			Blue Green		
Under Tail	White			White		

Breeding Period & Life Expectancy

Breeding Maturity

Gouldian finches reach breeding maturity between 8 to 12 months of age, at which point they also acquire their adult plumage. However, experienced breeders typically wait until the birds are in their second season before breeding them. This practice increases the likelihood of successful breeding by reducing the risks of egg binding and chick-tossing.

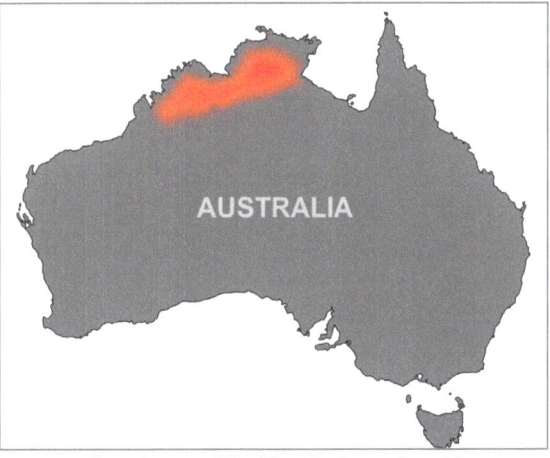

Distribution of Gouldian finches in the wild

Native Habitat Temperature Range - °C

		JAN	FEB	MAR	APR	MAY	JUN	JUL	AUG	SEP	OCT	NOV	DEC
KUNANURRA	Mean Maximum	36.4	35.5	35.4	35.5	32.9	30.5	30.3	33.6	36.4	38.3	38.8	38
	Highest Temp	43.9	42.2	40.5	39.5	38.0	36.7	36.4	39.4	41.0	43.6	45.1	44.6
	Mean Minimum	25.2	24.9	24.1	21.4	19.1	15.9	15.0	17.4	20.8	23.7	25.4	25.7
	Lowest Temp	19.6	19.0	16.1	12.2	9.6	7.7	4.8	9.2	10.6	15.0	17.2	19.1
WYNDHAM	Mean Maximum	36.9	35.9	36.0	35.8	33.5	31.0	31.2	33.5	36.7	38.8	39.4	38.1
	Highest Temp	44.7	44.0	43.6	41.6	38.8	36.9	37.6	39.6	41.7	45.0	46.0	45.4
	Mean Minimum	26.3	25.8	25.4	23.6	20.6	17.5	16.9	18.7	22.7	25.7	27.1	27.0
	Lowest Temp	20.1	21.1	19.0	15.7	10.5	8.3	9.0	10.4	13.5	17.6	18.6	18.6

See "References Used in This Book" on page 236

Hen Fertility

Hens are fertile for breeding up to 4 or 5 years of age, while cocks remain capable of breeding throughout their lifetimes.

Life Expectancy

In the wild, Gouldian finches typically have a life expectancy of 4 years (Red-Headed) to almost 6 years (Black Headed). However, when kept in an aviary environment with controlled conditions, such as protection from climate events, sufficient food supply and absence of predators, their life-span commonly extends to 7 to 9 years.

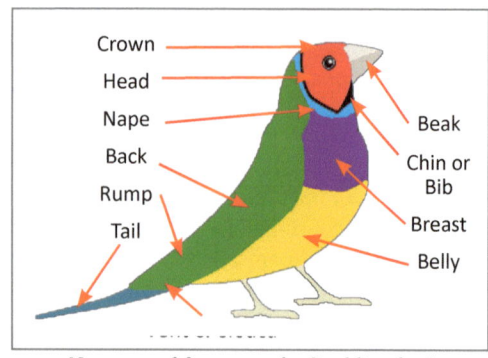

Names used for parts of a Gouldian finch

Average Body Size

Adult Gouldian finches weigh an average of about 15g (0.6oz), ranging from 12g to 18g. The cocks are about 135mm (5.3 inches) long, while the hens are 130mm (5 inches) having a shorter tail.

Basic Genetic Concepts

Finches have 39 pairs of chromosomes, with one of these pairs being the sex chromosomes. Since the development of DNA mapping, it has been shown that there is an enormous amount of information contained in the genes, or "genotype", of each bird. For example, the DNA sequence for a Zebra finch contains 1.2 billion individual

Gouldian finches are flocking birds. When one bathes, others often join in

letters.

It is therefore unsurprising that copying from one generation to the next can result in errors, thus there is the possibility of producing mutations - as discussed later in this book.

Terminology

As someone who cares for Gouldian finches it makes sense to use common terminology. This is better for understanding what others are referring to; and the best words to use are naturally the correct words.

As part of this introduction, a list has been prepared to assist in avoiding some of the most commonly mistaken words in Gouldian aviculture.

Behaviour Related to Head Colours

Gouldian finches not only exhibit three different head colours in nature, each of them also has associated behaviour patterns and mating preferences. Understanding these differences is also relevant in an aviary environment.

Red-Headed Behaviour

Red-headed Gouldian finches show more dominant and aggressive behaviour compared to their black and yellow-headed counterparts. Studies indicate that red-headed males tend to be more assertive in territorial disputes, nest selection and are more likely to engage in physical confrontations. This behaviour is associated with higher testosterone levels and affects their interactions with potential mates. It often makes them more successful in competitive environments, but they also experience higher stress levels and have shorter life spans.

Black-Headed Behaviour

Black-headed Gouldian finches are generally more reserved and less aggressive. They often exhibit cautious behaviour, which can help them avoid conflicts and predators. In social situations, they tend to be more submissive and can succeed in stable,

Terminology Mistakes	
Common Mistake	Correct Terminology
• Appearance	Phenotype
• Babies	Brood Chicks Nestlings
• Chest	Breast
• Children	Progeny
• Collar	Nape
• Egg Number	Clutch Size
• Euro Yellow	Yellow Back
• Female	Hen
• Genetics	Genotype
• Goldian	Gouldian
• Goldie	Gouldian
• Hatching	Incubating
• Home	Nest Box
• Jumped	Fledged
• Male	Cock Cock-bird
• Mouth	Beak
• Neck Ring	Nape
• Normal Chromosomes	Autosomes
• Pied	Australian Variegated Blue (AVB)
• Rainbow Finch	Gouldian Finch
• Teenagers	Juveniles
• Throat	Bib Chin
• Young	Fledglings

noncompetitive environments, resulting in longer life spans. Although the Red-Headed gene is dominant over the Black-Headed gene, black-headed Gouldians are more numerous in nature. This is due to their longer life expectancy and a natural mating preference for matching head colours.

Yellow-Headed Behaviour

Yellow-headed Gouldian finches have been observed to exhibit behaviours seen in both red and black-headed finches. They display moderate levels of aggression and caution, which allows them to adapt to different social environments.

(See "References Used in This Book" on page 236).

Getting Started

If you are planning to keep Gouldian finches, there are a few essential steps to ensure a smooth start. First and foremost, it is important to understand the specific needs of these birds.

One of the initial considerations is understanding their dietary requirements, followed by considering the type of environment you will provide for them.

Identifying Healthy Birds

When selecting Gouldian finches, it is important to choose birds that exhibit signs of good health.

Key Indicators to Look For

- Activity Level: Healthy Gouldian finches are active and alert. They will often be seen moving around, interacting with other birds and showing a general curiosity about their environment.
- Appearance of Eyes: Bright, clear eyes are a strong indicator of good health. Avoid birds with dull or cloudy eyes, as these can be signs of illness.
- Feather Condition: The feathers of a healthy finch should be smooth and vibrant in colour. Look for birds with well-groomed plumage without any bald spots or broken feathers.
- Behaviour: Healthy finches do not sit puffed up or display lethargy. Instead, they will appear energetic and engaged in their surroundings.

> **TIP** Always look for birds that are alert and interested in their environment.

- Signs of Illness: Avoid birds that look distressed, such as those that are sitting puffed up, have laboured breathing, or are showing signs of lethargy.

By carefully observing these characteristics, you can select healthy Gouldians and ensure a good start to caring for these beautiful and engaging birds.

Sexing Gouldian Finches

Determining whether an adult Gouldian is a cock or a hen (male or female) is quite easy for the natural wild-type birds. However, this can become more difficult for some of the mutations.

Natural Wild-Types

The wild-type forms are clearly distinguished for their sex by the darker colours of the males. The cocks have a purple breast and the hens are more of a lilac shade. However the breast colour changes for some of the mutations like White-Breasted and Yellow Back, but fortunately there are other signs to look for:

Illustration of the visible differences for sexing Gouldians

- Cocks have a darker shade of purple on the breast.
- Cocks have a more pronounced and more defined blue ring around the nape of the head.
- Cocks have a stronger yellow on their belly.
- Cocks have a sharper line between the belly and breast.
- Cocks have a stronger and more pronounced blue area at the base of the back.
- Cocks have a longer tail.
- Cocks perform an elaborate song.
- Cocks perform a courting display.

Example of sexing male and female Red-Headed Gouldians

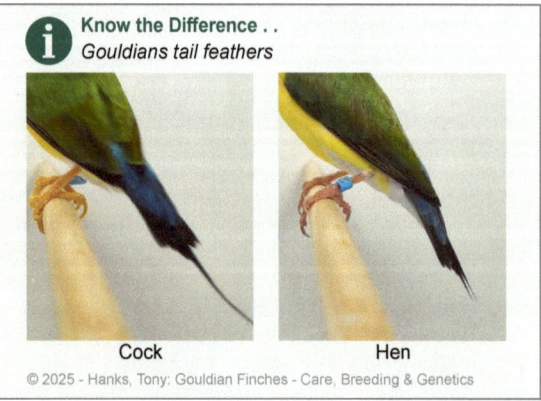
Example of the longer pin-tail seen in male Gouldians

Mutations

Some of the Gouldian mutations can be more difficult to sex, but the above list will still be helpful. The mutations themselves are discussed in the chapter about "Genetic Considerations". See page 119.

Quarantining New Birds

Introducing new finches to an aviary is an exciting time. However, it is crucial to quarantine the new birds before integrating them with the established flock to ensure the health and well-being of both the newcomers and the existing birds.

Benefits

Quarantining new finches is a precautionary measure designed to assess the health status of incoming birds; prevent the transmission of diseases to existing populations; and minimize stress for the new arrivals.

Period of Quarantine

An effective quarantine involves isolating the bird in a clean and disinfected separate enclosure away from other birds. The quarantine period for new finches is usually 14 to 21 days. This process requires patience, but it helps ensure the health and safety of the birds in the long term.

Gouldian Finches in Mixed Collections

Many people like the idea of a mixed collection of finches sharing the same aviary, however not all species are compatible with one another. Listed below are finch species that will generally create harmony in an aviary.

Recommended Species

- Chestnut Finches
- Cordon Bleu Waxbills
- Double-Bar Finches (Owl Finches or Bicheno Finches)
- Gouldian Finches
- Long-tail Finches or Masked Finches, but not both
- Painted Firetail Finches
- Plumhead Finches
- Red-Brow Finches
- Red-Faced Parrotfinch
- Society Finches (Bengalese)
- Spice Finches (Nutmeg Mannikins)
- Star Finches
- Zebra Finches (need to monitor, can be aggressive)

While these species are generally compatible in a mixed flight, it is important to observe their interactions closely. In particular, the following species should not be included with Gouldians.

Aggressive Species to Exclude

- Cut-Throat Finches
- Diamond Firetail Finches
- Java Sparrows
- Lavender Waxbills
- Melba Finches
- Saffron Finches

The aviary should have sufficient space for all birds to fly and seek refuge as needed. Providing multiple feeding stations and nesting sites can also help minimise competition and stress.

Essential Supplies and Equipment

To ensure the aviaries and breeding cabinets are well-equipped for Gouldian finches, several essential supplies and pieces of equipment are required. Feed and water are discussed later, but the other necessary items include:

Perches

Perches of varying diameters and textures help maintain healthy feet. It is important to place them at different heights and locations within the enclosure to encourage natural movement and exercise.

If you have a more protected area where you would prefer the birds to sleep at night, it is a good idea to position the perches higher in this area. This is because the birds feel more secure when roosting in elevated positions.

Nesting Boxes

Providing ample nesting boxes is crucial for breeding pairs. These should be clean, secure in their position and appropriately sized to give the birds a safe place to lay eggs and raise their young.

For Gouldian finches, an enclosed nest size of 17L x 14W x 16H cms (7 x 5.5 x 6 inches) is a good guideline. These are discussed further in the chapter titled "Breeding Gouldian Finches". See page 95.

To avoid territorial fighting in an aviary or flight there should be more nest boxes than breeding pairs of birds. For example, three pairs in a flight should be offered at least 4 nesting boxes.

Gouldian Nest Box

Food Dispensers

Ensuring access to hygienic food dispensers is essential. Some individuals utilize an open bowl for seeds, while others prefer a hopper. Open bowls replicate the natural foraging conditions, but they necessitate daily removal of husks to maintain access to the seeds.

Seed hoppers require less

Seed hoppers are convenient

frequent refilling; however, they are prone to clogging if oily supplements have been added to the seeds. Both types of dispensers should be cleaned regularly to prevent contamination and guarantee that birds consistently have access to fresh seeds. Detailed discussions on this topic can be found in the chapter on "Nutrition and Feeding".

Water Dispensers

Accessible and hygienic water dispensers are essential for providing water to birds. Open dishes allow for both drinking and bathing, while hoppers help keep the water clean. An irrigation timer can refill the bowl with fresh water regularly when open dishes are used.

Alternatively, an Edstrom water valve releases a drop of water activated by the bird's beak. These can be gravity-fed with micro-irrigation pipes or attached to individual bottles. Regardless of the system, it should be cleaned regularly to prevent contamination and ensure that birds have continuous access to fresh water.

Automatic filling for a water bowl

Bathing Facilities

Regular bathing is essential for plumage maintenance. So, if an open bowl is not already used for water, a shallow dish or bird bath should be placed in the enclosure every week to allow finches to bathe and keep their feathers in top condition.

Edstrom water valves

Cuttlefish holder

Providing cuttlebone is essential for the calcium intake of finches. These natural sources of calcium help in the development of strong bones and eggshell formation in breeding females. Position the cuttlefish holder in an easily accessible area within the enclosure and ensure it is securely attached to prevent it from being knocked over. Regularly check the cuttlebone for any signs of mould or contamination and replace it as necessary.

Shell Grit Container

Shell grit is an important supplement for finches, providing essential minerals, especially calcium, which is crucial for egg formation and overall health. The grit aids in digestion and helps grind down food in the bird's gizzard. It is best be offered in a separate dish or coop cup as part of their standard diet. Regularly check the shell grit for freshness and contamination, replacing it as needed to ensure it remains a clean and reliable source of nutrients for the birds.

Nesting Material

Finches prefer a variety of nesting materials that are soft and pliable, enabling them to build a comfortable and secure nest. Common materials include coconut fibre, dried grass, hay, sisal, cotton wool and shredded tissue paper. It is important to avoid synthetic fibres that could entangle the birds or pose a health risk.

Providing an assortment of these materials will encourage natural nesting behaviour and ensure that the finches can choose what best suits their needs. Place the nesting materials in an accessible container within the flight or cabinet, then regularly replenish the supply to keep the birds engaged and to maintain a hygienic nesting environment.

Sisal is a convenient option for nesting material

Catching Net

Those with flights will eventually need a net to catch their birds for relocation. Even those with only breeding cabinets will find a net is needed when a bird escapes into the bird room.

A catching net is a necessary for safely capturing and handling finches without causing them undue stress or harm. When selecting a catching net, one with a soft fine mesh is better than a more open net that may cause injury to the delicate feathers and limbs of

Open netting on a catching net can cause injuries, so models with fine netting are preferred

the birds. The rim should be padded and the handle should be long enough to allow you to reach birds in larger enclosures with ease, but lightweight enough to manoeuvre swiftly and precisely. In smaller flights a larger net can be too difficult to handle, so a 30cm (12 inch) diameter is a good compromise.

Transport Cage

Right from the beginning of keeping and caring for Gouldians, another piece of necessary equipment is the transport cage. This will be used for safely moving the birds, whether moving them to a new permanent home, veterinary visits, shows, or. These cages should be sturdy yet lightweight, providing adequate ventilation while preventing the birds from escaping or injuring themselves during transport.

A well-designed transport cage should have a secure locking mechanism, easy access for placing birds with a rubber safety shield, a simple release door with no further handling, and a comfortable perch.

BASIC GETTING STARTED CHECKLIST	
❏ Housing - protected & secure	See page 31
❏ Mixed collection or only Gouldians?	See page 23
❏ Perches and cuttlebone holder	See page 24
❏ Seed container - bowl or hopper?	See page 69
❏ Fresh water - bowl, hopper or valve?	See page 70
❏ Shell grit container	See page 25
❏ Catching net if housing is a flight?	See page 26
❏ Transport cage to bring the birds home	See page 27
❏ Purchase seed - Finch Mix	See page 51
❏ Purchase supplements	See page 55
❏ Purchase shell grit, cuttlebone	See page 56
❏ Birds - healthy and not over-crowded	See page 32
❏ Release into the flight or cabinet	

Transport cage

During transport it is important to have seed available to maintain the birds' metabolisms, but water is not necessary for trips less than 2 hours. The transport cage must be thoroughly cleaned and disinfected after each use to prevent the spread of disease.

Hospital Cage

These special isolation cages provide a safe and controlled environment for sick, injured or recovering birds. The bird can be closely monitored and treated without stress and any infection can be isolated from other birds. Further information is discussed in the chapter about "Health & Wellness". See "Use of a Hospital Cage" on page 88.

Stress Perches

Stress perches are perches that are made with a number of individual areas, each one for only one or two birds. This can help in preventing behaviours such as feather plucking or aggression, therefore promoting natural, healthy behaviours.

Climate & Temperature

Stress perches

Gouldian finches thrive in a setting that closely mimics their natural habitat. This means maintaining a warm climate with temperatures ranging between 20-30°C (68-86°F) and ensuring proper humidity levels of around 50-60%.

Minimum Temperatures

Temperatures below 15°C (59°F) are very cold for Gouldian finches. This is because, as a tropical bird, they do not have a layer of down between their feathers and skin. Down feathers are typically used by birds for insulation, trapping air and maintaining body heat. So, this absence of down makes Gouldian finches particularly sensitive to cold weather and cool draughts.

> **TIP** Gouldians are more tolerant to higher temperatures, but they are very susceptible to cold draughts.

Some carers claim that colder conditions will "harden" the birds to these conditions, but they are simply not suited to continued cold weather and health problems will develop as a result. For these reasons most carers provide artificial heating to maintain a stable, warm environment when conditions are too cold.

Maximum Temperatures

Temperatures above 40°C (104°F) are not uncommon in the natural habitat of Gouldian finches, however this extreme is bordering on the range where incubating eggs will die, or the chicks will be born crippled or deformed.

High temperatures can be quite common under transparent polycarbonate roofing, so some form of insulation may be necessary. Another popular option for a flight or bird room is an extraction fan controlled by a thermostat above 30°C (86°F).

Leg Rings and Record Keeping for Birds

Leg rings are an essential tool for breeders to accurately keep track of individual Gouldian finches and their lineage. By assigning a unique colour and number combination to each bird, breeders can identify and monitor the health, breeding history, and genetics of their flock.

Ring Size

There is a wide range of leg ring sizes available, so it is important to purchase an appropriate size for Gouldians. An opening size of between 2.2 and 2.8mm will fit comfortably on Gouldian legs.

> **TIP** A leg ring diameter of 2.5mm is the recommendation for normal use on Gouldian finches.

Split or Fixed

Split leg rings can be fitted to the bird's leg at any time, including when a new adult is purchased. Fixed rings are only fitted to birds in the nest when they are able to slip over the bird's foot before it grows.

Of these two options, split leg rings are the easiest to use since they don't require disturbing chicks or handling small birds. They are usually applied when juveniles are caught for removal from their parents.

Split leg rings are supplied with an applicator tool used to open the ring and slide it onto the bird's leg.

Split plastic leg rings are supplied with an applicator

It is recommended to consistently apply them to the same leg (based on whether the person is right or left-handed) and to position the top of the number closest to the bird's body. This involves placing the ring on the applicator with the top of the number nearest the tip of the tool. Consistent application prevents confusion between similar numbers, such as 6 and 9.

Accurate Records

Accurate record-keeping is essential when selecting pairs for breeding in a new season to prevent inbreeding and aim for desired outcomes such as improved health, adherence to show standards, or specific colour combinations.

Some aviculturists maintain detailed records of their birds and breeding activities in books, while others utilize specialized software for this purpose. See an "Example of a Computer Based Bird Record" on page 120.

An example of a paper record is shown on the following page and a resource for maintaining these is included in this book. The form can then be photocopied as needed to create a personal logbook. See the "Forms Library" on page 203.

Bird Log for Gouldian Finches

Ref Number	26		Leg Ring	Red 07	Season Yr	2024
Description	Yellow headed cock				Date Born	1/7/2024
			Sex	☑ Ck ☐ Hn		☑ Estimate

Phenotype:

Head		Tip of Beak		Date Died	
☐ Red	☐ Grey	☐ Red	☐ White		
☐ Black	☐ Lt Grey	☑ Yellow		**Purchased**	
☑ Yellow	☐ Lt Salmon	**Belly**		Date	15/1/2025
	☐ Salmon	☑ Yellow	☐ White	From	J Wilkes
	☐ White	☐ Lt Yellow		Amount	$ 30

Breast		Back			
☑ Purple	☐ Lilac	☑ Green	☐ Pastel Blue	**Sold**	
☐ White		☐ Yellow	☐ White Silv	Date	
		☐ Yellow Mrks	☐ Seagreen	To	
		☐ Dilute Grn	☐ AVB	Amount	$
		☐ Blue	☐		

Genotype:

Sex Linked						
Z	Z	W	Z	Z	Z	Z
☑R ☐b	☑R ☐b	☒Hen	☐ YB	☐ YB	☐ cn ☐ sg	☐ cn ☐ sg

Autosomes				Genetics Note	
☑ yh ☑ wh	☐ wb ☐ wb	☐ bk ☐ bk	☐ ay ☐ ay		

Date	General Notes, Pairings, Health, Medications, etc
22/1/2025	Moxidectin for air sac mites
10/2/2025	Paired with Yellow-Headed hen ref no 16

Example of a paper based Bird Record

3

Housing & Protection

Every individual responsible for Gouldian finches has a duty to provide appropriate housing for these birds. Proper housing is essential for maintaining their health and well-being. Without suitable living conditions, Gouldian finches are at a significant risk of not surviving.

Choosing an Aviary, Flight, Cage, or Cabinet

When deciding between an aviary, flight, cage, or cabinet, it is necessary to consider the number of birds you plan to keep and the space you have available.

The Housing Options

We can make the following observations about these choices:

- Aviaries are ideal for larger groups and allow for more natural movement and social interaction. The birds have plenty of room to fly and they can also be planted with grasses and trees.
- Flights are effectively a division of a larger aviary in order to keep a specific variety of birds together. For example, a particular head colour or variety of Gouldians.
- Cages designed for canaries or budgerigars are not suitable for Gouldian finches due to their small size, exposure to cold draughts and not providing a sense of security for the birds.
- Cabinets are designed for breeding pairs of Gouldians. They are typically larger than cages and open only on one side to provide protection for the birds.

> **TIP** Never use an all wire birdcage for Gouldians because these enclosures are draughty and insecure..

One common approach is to use both cabinets and flights. The cabinets facilitate breeding of selected pairs, while the flights are used for raising juveniles and housing other birds during the non-breeding season.

Factors to Consider

All cages and flights should provide adequate space for exercise with sufficient horizontal space for flight, rather than just vertical height. Including natural branches, multiple perches and a foraging area on the floor will support natural behaviours and keep the finches active. It is also necessary to select a design that allows for regular cleaning to maintain hygiene and prevent disease.

An aviary flight

At some point, the caregiver will need to catch birds for relocation. An aviary or flight enclosure should be sufficiently large to allow entry and for the owner to stand comfortably. Breeding cabinets should be designed in a way that enables the birds to be caught without causing excessive stress or exhaustion. Implementing a removable dividing panel is a good solution for this purpose.

Breeding cabinets for Gouldians

Flight or Cabinet Sizes

Aviaries, flights and cabinets should always be as large as possible, but never smaller than the minimum size requirements for the number of birds to be housed. (See "Guidelines for Number of Birds" following).

As an example of sizes, a typical aviary or flight is 2.1M high, 3M long and 1.5M wide (7' x 10' x 5'). By comparison, a typical breeding cabinet is 0.6M high, 1M long and 0.4M deep (2' x 3'3" x 1'4").

Guideline for Number of Birds
Volume or Area

Space is often referred to in three-dimensional cubic metres or cubic feet. These are calculated by multiplying length x width x height. However, the

> **TIP** Never be tempted to overcrowd a cabinet or flight. Doing so can cause fighting, bullying, spread of disease and poor breeding results..

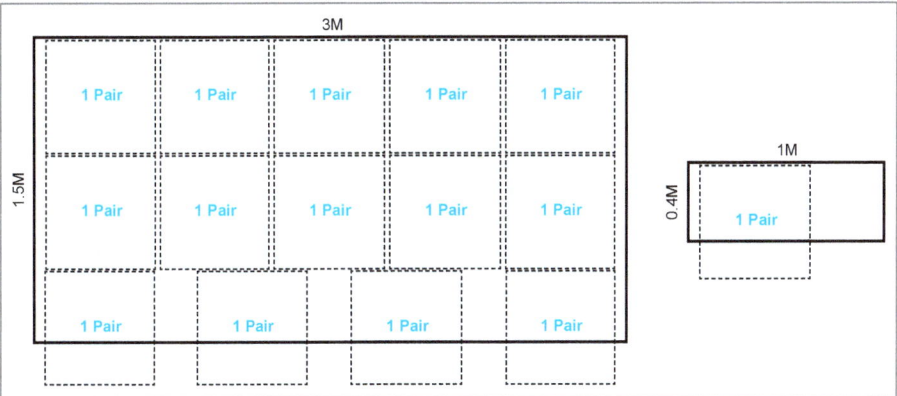
Visualisation of the housing guideline for flights and cabinets

height of the enclosure does not play an important role because finches make more use of the horizontal, rather than the vertical. They also prefer the highest perches, so a better guide is the two-dimensional area (length x width) expressed as square centimetres or square feet.

Housing Guideline

Therefore the housing guideline for Gouldian finches is . .

> One pair per 0.27 to 0.37 square metres (M^2)
> One pair per 3 to 4 square feet (ft^2)

As an example, a flight of the dimensions mentioned above will have a horizontal area of $4.5M^2$ (3 x 1.5) or $50ft^2$ (10 x 5); so this could accommodate approximately 14 pairs of birds.

A breeding cabinet of the dimensions mentioned above has a horizontal area of 0.4M2 (1 x 0.4) or 4.3ft2 (3'3" x 1'4"); so as expected, this comfortably accommodates one pair of birds.

These calculations allow sufficient space for each bird to fly, socialize, and engage in natural behaviours without feeling overcrowded. Overcrowding can lead to stress, aggression and the spread of disease, so it is crucial to adhere to these guidelines to ensure the well-being of the birds.

Wind Protection

Danger of Cold Draughts

Cold winds can be a specific challenge to Gouldian finches, as they increase the difficulty of maintaining warmth in already low temperatures by removing any residual warmth that the birds might retain. These birds are highly susceptible to chilling winds or draughts, which can result in rapid heat loss, hypothermia and a higher risk of respiratory infections.

Wind Barriers

To reduce the adverse effects of cold winds, it is crucial to provide adequate shelter that shields the birds from draughts. Enclosures should be strategically placed away from the direct path of winds, and additional barriers like windbreaks can be useful in outdoor settings.

Wind protection in the form of shutters

Location & Aspect
Direction

A critical factor to consider is the location and orientation of the aviary, flight, or bird room. Gouldian finches thrive when exposed to natural sunlight and they must also be sheltered from the coldest winds. In the southern hemisphere, the best orientation is north-east to capture the morning sun, whereas in the northern hemisphere, the preferred direction is south-east.

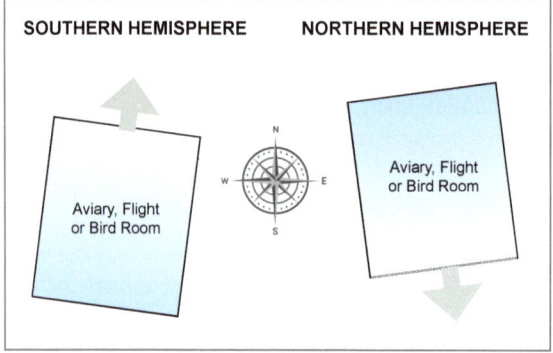
Direction facing aspects for Gouldian finches

For outdoor enclosures, adding weatherproof covers and windbreaks can offer additional protection.

For indoor enclosures, choose a quiet location that is away from draughts or sudden temperature changes. Do not place cages near air conditioning or heating vents.

Natural Sunlight

Natural light has an important role for Gouldian finches. This is discussed under "Sunlight & UV Light" on page 35.

Cold Weather Protection

Gouldian finches are susceptible to cold weather and require proper insulation to maintain warmth. Materials such as wood or plastic can be used to create a barrier against cold winds and to ensure that the enclosure is free of draughts.

Depending on the minimum temperatures in their location, some breeders may add a safe heat source. Suitable options include ceramic heaters or column oil heaters.

> **TIP** Positioning is critical to the success. Gouldians must be protected from cold winds and they must also have access to natural light.

Hot Weather Protection

In hot weather, it is important to provide shade and ventilation to prevent overheating. Nests placed near the roof can be a particular problem because this area tends to be the hottest part of the enclosure.

Ventilated roofs for air circulation; a double roof, roof insulation; or locating the enclosure in a shaded area can help keep the birds cool during summer. In some hotter regions, regularly misting the area with water can also help to provide relief from the heat.

Sunlight & UV Light

Sunlight is essential for maintaining the health of birds, similar to its importance for many other creatures. It offers critical benefits that enhance their physiological and psychological conditions.

Vitamin D Synthesis

Sunlight is a primary source of ultraviolet (UV) light, which birds need to synthesize Vitamin D3. This vitamin then plays a critical role in calcium metabolism, influencing bone health, eggshell strength and overall vitality. Without sufficient Vitamin D3, birds can suffer from brittle bones and other calcium deficiency-related issues such as feather loss.

Regulating Circadian Rhythms

Exposure to natural light plays a crucial role in regulating birds' circadian rhythms, which are essential for their natural day-night cycles. These rhythms are needed for sustaining healthy sleep patterns, feeding behaviours and overall activity levels. Any disruption of these rhythms can result in stress, anxiety, and behavioural problems.

Mental Stimulation and Well-being

Natural sunlight provides mental stimulation for birds. The changing light patterns throughout the day mimic their natural habitat, contributing to their overall well-being and reducing the likelihood of depression and lethargy. Sunlight also improves the brightness of their plumage, which plays an essential role in their visual communication.

Full-Spectrum Lighting in a Flight or Cabinet

While having a wall with direct access to natural sunlight is one solution, it may not always be sufficient or practical. It is also important to ensure that there is a balance of light and shade, similar to the conditions found in nature, rather than exposing them to constant direct sunlight.

When natural sunlight is inadequate, full-spectrum artificial lighting can be used. These lights replicate the ultraviolet (UV) spectrum, and UVB bulbs are commercially available. However, these UV lights can be challenging to purchase, as they have been banned in many countries due to the potential risk of skin cancer in humans.

Recommendation

Adequate sunlight or its equivalent is necessary for the health of all birds. LED lights designed for plant growth can offer full spectrum lighting, enabling Gouldian finches to see a wide range of colours. However, these lights do not provide UVB light as would be found in natural sunlight.

When natural sunlight is unavailable or insufficient, it is recommended to use LED "Growth Lights" along with a supplement that includes Vitamin D.

LED Growth Lights will still need a Vitamin D supplement

Cleaning and Maintenance

This topic is discussed in the chapter about "Health & Wellness" (see page 74), but it should be considered and planned for at the time of designing flights and breeding cabinets. Cleaning is essential for preventing disease, so it must be easy to perform. Maintaining a dry floor is also important for avoiding diseases like coccidiosis and again it depends largely on the chosen design.

Floors of Aviaries & Flights

When designing an aviary, selecting the appropriate flooring is important for maintaining hygiene and ensuring the well-being of the birds. Several flooring options exist, each presenting its own benefits and challenges:

Concrete Floors

Concrete is commonly selected for its durability and ease of maintenance. It offers an effective barrier against pests, including rodents and insects. However, it can become extremely cold during winter, particularly if is not well drained.

Natural Earth Floors

Using natural earth or soil as flooring can create an environment for birds to forage and display natural behaviours. However, this type of flooring requires more maintenance to prevent it from becoming muddy or infested with rodents. It may also produce an odour if it remains too wet.

Sand Floors

Sand is an option that replicates the birds' natural environment. It is easy to clean and can be replaced periodically to maintain hygiene. Sand also assists with drainage, reducing the risk of a damp setting. However, it can sometimes harbour parasites and may require frequent changes.

Wooden Floors

Wooden floors offer natural and effective insulation properties. They can be layered with materials like pine shavings or shredded paper to absorb moisture and droppings. However, wood is susceptible to rot and may need regular treatment with bird-safe preservatives to maintain its life-span.

Wire Mesh Floors
Wire mesh on suspended flights keeps the area clean by letting droppings and debris fall through, but it can harm birds' feet and cause injuries. Proper maintenance and providing perches can help prevent these issues.

Gravel Floors
Gravel provides excellent drainage and a natural appearance. It also assists in maintaining the birds' nails at an appropriate length. The gravel needs to be regularly raked and cleaned to prevent the accumulation of bacteria.

Gravel floor

Recommendation
A gravel layer on a concrete floor offers the best compromise for aviaries and flights. The concrete ensures that the floor is vermin-proof and the gravel provides the drainage to keep the surface as dry as possible.

Floors of Breeding Cabinets
When considering the ideal flooring for bird cages, it is again essential to balance hygiene, maintenance, and the birds' comfort. Most breeding cabinets have a removable tray, but there are various options for the lining that becomes the floor for the birds:

Sand Filling
Sand floors provide a natural environment for birds, which can enhance their well-being. These floors are easy to clean and can be periodically replaced to maintain hygiene. Sand also assists with drainage, thereby reducing the risk of damp conditions. However, water can sometimes become trapped by the tray and the sand then becomes soaked. Sand can also harbour parasites, requiring frequent changes to ensure a clean and safe environment.

Paper Lining
A disposable paper lining, such as newspaper or brown paper, is commonly used for breeding cabinets. These materials are straightforward to remove and replace, but may require frequent changes due to droppings accumulating below the perches. Additionally, paper can become a problem if it gets saturated and adheres to the tray, complicating the cleaning process.

Pine Shavings

Materials like pine shavings or shredded paper help to absorb moisture and droppings, contributing to a cleaner environment. Pine shavings also have a pleasant odour.

Animal Litter

Another alternative is commercially available animal litter, typically marketed for cats and various small animals such as rats and reptiles. While

Pine shavings in a breeding cabinet

these products generally offer excellent moisture absorption, it is important to note that some may be treated with chemicals that could make them unsuitable for use with finches.

Wire Mesh Floors

Wire mesh floors are commonly utilised in a two-tray system above the dirt tray beneath. These floors help maintain cleanliness in the cabinet by allowing debris to fall through; however, some droppings may still accumulate on the wire. While effective, wire mesh can pose a risk of injury to birds' feet and can also be difficult to clean.

Recommendation

Pine shavings offer a good choice for hygiene, ease of maintenance and the birds' safety. This should also be combined with a design that does not direct water spillage onto the cabinet floor.

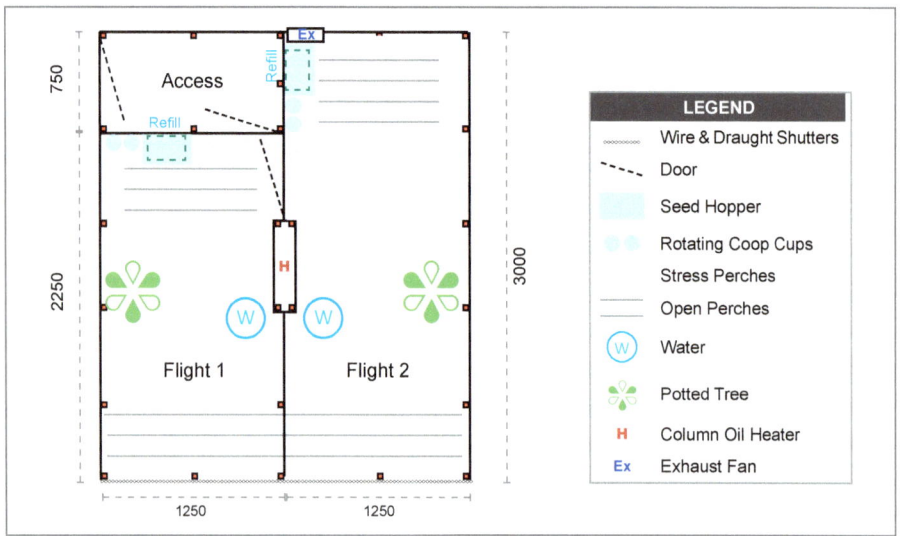

Example of 2 flights that are designed for Gouldian finches

Custom Design for a Flight

The aim of building an aviary or flight is to provides a spacious environment for your birds and the steps involved are as follows:

Planning and Design

The purpose of a flight for Gouldian finches will either be for breeding in colonies; for holding independent juveniles until they moult; or for separating cocks and hens in the non-breeding season. In most cases it will be for more than one of these purposes.

The location was discussed earlier and it will need ample sunlight, shade and protection from harsh weather. Also consider adding a double door system to prevent birds from escaping when entering or exiting each flight.

The design will then be based upon the purpose, location, amount of space available and the budget.

The finished flights with wind shutters installed

Building Materials

The choice about building materials will depend upon many factors, including the climate of the location and the ability of those doing the construction.

A popular choice for the frame is either steel or timber, however in coastal locations aluminium is a better choice. Roofing can be metal or polycarbonate. The acrylic choice will give better light level results, however it can also create issues with excessive heat in summer.

Wire mesh will be required for some walls and here there are also many choices available. Decisions need to consider the goal of vermin exclusion, so 6mm vermin wire is better in that regard. It's also a benefit to use black coated wire so that it's easier to view the birds.

Construction

The first decision for the owner is to decide whether they will need to engage a builder or enjoy doing the construction themselves?

Essentially the steps involved are to pour a concrete floor; build the frame using wood or

An inside view showing perches and stress perches

metal; attach the wire mesh; install the roof; attach the wall linings; and install doors.

Interior Setup

The final step is to add the aviary equipment like perches, feeding stations and nest brackets. For perches they can be natural branches or wooden dowel. When positioning remember that birds generally prefer to sleep in a high location, so if you'd prefer the birds to sleep in the more secure end of a flight it is advisable for these perches to be the highest available.

Potted tree with drainage below to keep the flight dry

Feeding stations will also need to be in areas that are protected from weather, easy to access for refilling and not directly below perches where they could be fouled by bird droppings.

Example of a Flight Design

The "example of a design for 2 flights" on page 38 was specifically designed for Gouldian finches. This design therefore included the following features:

- North easterly aspect (Southern hemisphere) to receive winter sun and avoid cold winds
- Aluminium frame for rust avoidance because the location was coastal
- Nylon aluminium tube connectors with enclosed steel core because of potential wind exposure
- Polycarbonate roofing in an Opal White colour to balance excellent light transmission without excessive heat
- Vermin wire to exclude mice, including under the polycarbonate roofing
- Concrete floor for better drainage and to exclude vermin
- Open wire limited to only one surface to protect form cold draughts
- Access area with a second door to prevent accidental escapes
- Automatic water system that includes drainage to take outside the flight
- Integrated feeding stations that can be refilled from outside the flights
- Perch and nest locations that avoid droppings on feed or water below
- Design includes a tree growing in a pot that can be removed and replaced if necessary
- Ventilation on a thermostat for hot summer days
- Heating on a thermostat for cold summer nights
- Gravel floor above the concrete base for good drainage and a dry environment

Housing & Protection

A list of the building materials is included in the "Appendices" of this book (see "Products and Suppliers" on page 232).

Custom Design for a Breeding Cabinet

Parts of the following discussion were first published by the author in the "Australian Birdkeeper" magazine, April-May 2025 issue.

Gouldian finches can be bred successfully in aviary colonies, but achieving a specific genetic outcome requires using cages for individual pairs.

One of the most crucial aspects of genetics-controlled finch breeding is providing an optimal living environment, which starts with a custom-designed breeding cabinet. These cages are tailored to meet the specific needs of finches, ensuring their health, happiness and successful breeding.

When designing a custom breeding cabinet, it is essential to consider factors such as space, materials, ventilation and accessibility.

TIP When building or buying an enclosure, always use wire that is coated black for better visibility.

Space Requirements

Finches require ample space to live, fly, and breed. A custom-designed cage should allow for horizontal flight, as finches are not strong vertical fliers. A breeding cabinet should be spacious enough to accommodate a single pair of finches, together with up to 5 juveniles until they become old enough to separate.

The dimensions of the breeding cages are 100 cm in length, 44 cm in depth and 53 cm in height. Although larger sizes are always preferred, these dimensions were chosen

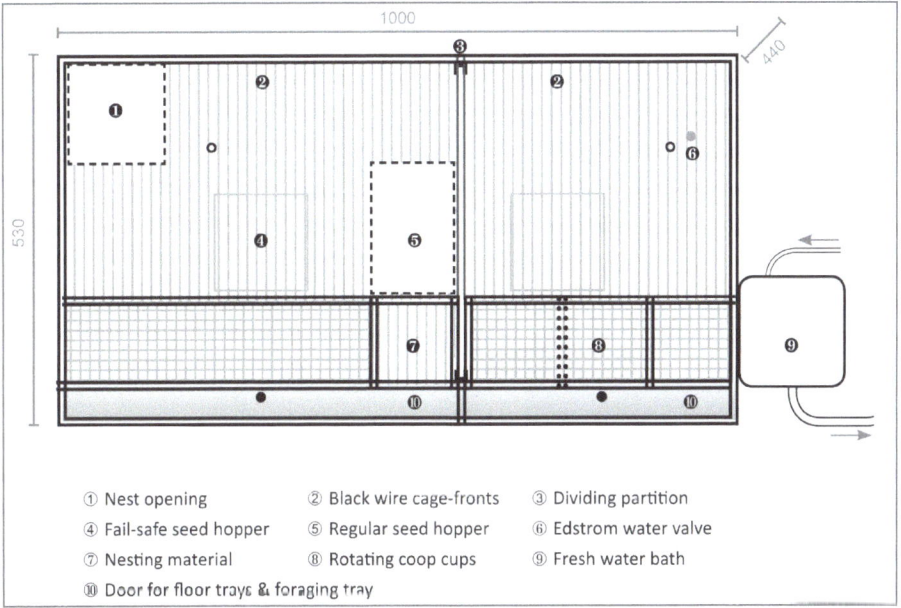

① Nest opening ② Black wire cage-fronts ③ Dividing partition
④ Fail-safe seed hopper ⑤ Regular seed hopper ⑥ Edstrom water valve
⑦ Nesting material ⑧ Rotating coop cups ⑨ Fresh water bath
⑩ Door for floor trays & foraging tray

to optimise the available space while allowing for the vertical arrangement of three cages high.

Materials and Construction

The materials used in constructing a bird cage are important for the birds' safety and well-being. Non-toxic, durable materials such as powder-coated metal wire, aluminium frames, and aluminium composite panels (ACP) are ideal choices. These materials are easy to clean and resistant to rust and corrosion, contributing to the long life of the cage.

The bars of the selected cage fronts were spaced at 10mm, which is more suitable for finches compared to the standard 12mm (half inch) separation. These are pre-coated black to reduce maintenance and improve visibility.

Aluminium tube of 20mm square is an ideal building material for the frames. They were assembled using nylon joiners in a range of options, including elbows, tee-joints, 4 and 5 way joiners.

Aluminium composite panels are frequently used for sign-writing and are easy to work with. They are available in various colour options. However, gloss white, while effective on the exterior, was found to be unsuitable for the interior surface of cages due to the reflections causing stress for the birds' predator awareness. Therefore, this surface was painted matt white.

Finished breeding cabinets are arranged three units high

Ventilation and Lighting

Proper ventilation is necessary to maintain a suitable environment for breeding finches. The cage should allow adequate airflow to keep the air fresh, while also preventing cold draughts for Gouldians. This can be achieved with an open cage-front on one side and closed panels on the other three walls, floor, and ceiling.

Additionally, providing natural or full-spectrum lighting is important for birds' overall health and breeding success. Modern polycarbonate roofing allows adequate lighting, but it blocks 100% of the UV spectrum, which is essential for bird health. This issue can be addressed by installing full-spectrum lights that mimic natural sunlight during normal daylight hours and help regulate the birds' circadian rhythms. These lights also enable birds to see the full range of colours, promoting better breeding conditions.

Accessibility and Maintenance

A well-designed bird cage should be easy to access for both the birds and the caretaker. Large doors and removable trays make cleaning and maintenance more convenient. Incorporating multiple perches, feeding stations, and water dispensers at different heights encourages natural behaviours. Nesting boxes should be easily accessible for breeding pairs, allowing them to build nests and lay eggs in a safe and comfortable environment.

When designing these custom breeding cabinets for Gouldian finches, several additional features were included to enhance the birds' well-being, simplify their care and improve breeding success rates.

Water Options

In all aspects of breeding Gouldian Finches, I employ a "Fail Safe" approach. This is particularly crucial for both seed and water. Two

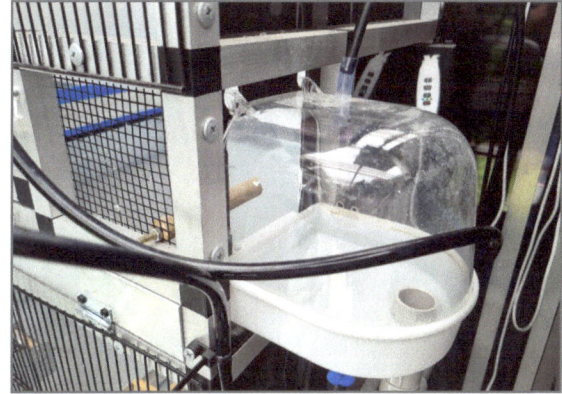

Commercial water baths with water input and overflow drain

completely independent systems supply fresh water to the cabinets. Should one system fail (for instance, due to a flat battery), the second water source remains unaffected and continues to function without interruption.

The first option is gravity-fed Edstrom water valves. Birds adapt to this system quickly, similar to how they drink water off the bird wire after rainfall.

The second water source is an externally mounted bird bath. It operates with mains pressure water, controlled by a timer and runs for 2 minutes daily. The system includes a central drain that allows the bath to be flushed clean with fresh water each day.

Feed Options

Finches have a rapid metabolism and need to eat about six times a day. Access to seed is essential, so a "Fail Safe" approach was taken. The main source of seeds is a hopper with enough capacity to last up to 2 weeks. Additionally, there are coop cups and a designated foraging area on the floor. When going away, a second hopper is installed into one of the cage doors as an extra precaution.

Previous experiments featured attractive plastic hoppers, however these were not well-received by the birds. The design required that they enter the hopper to access the feed, which caused discomfort and heightened caution due to their

Final choice of hoppers presents the seed inside the cage

natural wariness of potential predators. Instead, metal hoppers with an extended tray into the cage have been significantly better accepted.

Rotating Coop Cups

Stainless steel Coop Cups are well-suited for holding shell grit, baked eggshells, sprouted seeds and chopped vegetables. To enable easy refilling with minimal disturbance, a rotating design was used in the breeding cabinets. (Using a drill press during fabrication significantly improved the positioning accuracy of the vertical rotation rods within the cages).

"Fail Safe" hoppers are in place during absences of the owner

Floor Trays

The initial plan for the breeding cabinets involved a two-tray system on the floor: A lower "Dirt Tray" with a "Wire Tray" above it. Although these were constructed, the wire trays were challenging to clean, leading to their replacement with a single-tray system.

The final single trays are 25mm deep and filled with fresh pine shavings. The pine absorbs moisture, has a pleasant smell, provides a natural surface for the birds, and is easy to clean by simply replacing it. A 40mm horizontal door is used in front of the tray to allow ample space when removing the tray for cleaning and so that the door can be re-closed during that time.

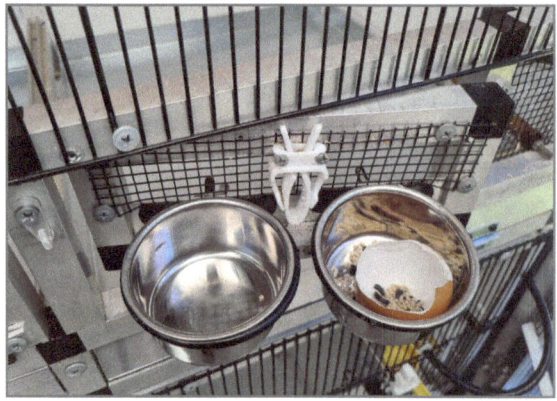
Rotating coop cups allow easy refilling without disturbance

Foraging Trays

Gouldian finches, as flocking birds, often feed and forage together. To allow for this

An access door opens to slide out the floor tray

natural behaviour each breeding cabinet contains a plastic foraging tray measuring approximately 18 by 22 cm. These trays are placed away from perches and are used to collect spilled seed and husks, crushed eggshells, seeding grasses, shell grit and charcoal.

Nesting Areas

Nesting areas within the cage are obviously essential for breeding finches. Nesting boxes should provide a sense of security and are more effective when designed specifically for Gouldians. These boxes can be constructed from untreated plywood with an external verandah leading to a lowered nest area. Both features aim to reduce chick-tossing, and a wire floor on the verandah helps to maintain cleanliness.

The wire cage fronts have doors to position the nest boxes, and blocks of timber were added to the boxes so that they lock securely into place due to their own weight.

The foraging tray in use by a pair with fledglings

Nests can be locked into place with their own weight

Nesting Material

Each nest box is installed with nesting material already in place, but birds often add more, or as part of their courting procedure. Nesting material is provided in a vertical holder to keep it clean before use. This arrangement also makes it easy to monitor when the new nesting material is being utilized.

Once the eggs are being brooded, additional nesting material is not provided. This measure ensures that overly enthusiastic cocks do not continue building on top of the eggs.

A Dry Floor

For finches, it is important to maintain a dry floor to reduce the risk of diseases such as coccidiosis. Although water valves are an effective system, birds often block them with seed husks from eating, which can result in continuous dripping onto the floor below. In addition to the potential health risks, a wet floor can also have an unpleasant odour.

The potential dripping problem was resolved by redirecting any water back outside the cabinets. This was achieved using the base of a water hopper, but reverse mounting it with the larger portion inside the cage and the spout outside. A tube is connected to deliver the water to the same drainage used for the water baths. It is important to note that this gravity-fed drain cannot have too small a diameter. (Refer to "Bernoulli's Principle" online for more information).

Finished cabinet ready for pine shavings on the floor

Catching Dividers

Breeding cabinets with a length of one metre can be inconvenient for catching finches. A sliding vertical divider solves this issue and is used to catch and remove juveniles once they become independent.

Temperature Control

Maintaining appropriate temperature and humidity levels is essential for breeding finches. Custom-designed cages can include features such as heating elements or humidifiers to regulate these factors, with ideal temperatures ranging between 20°C and 30°C and humidity levels around 50-60%. My preferred option is to achieve this environment in the bird room, rather than the individual breeding cabinets.

Two thermostats are used: one to activate a column heater if night temperatures fall below 12°C, and another to turn on an extraction wall fan when the temperature exceeds 35°C. Keeping the environment stable and within these ranges can enhance breeding success and overall health.

Supplementary Lighting

The significance of full spectrum lighting has been previously mentioned. Modern roofing materials limit the amount of UV light birds receive, which can also impact on Vitamin D absorption. Therefore, some form of supplement is necessary—either a dietary supplement or artificial light. Both of these methods are used for Gouldians under roofing, but this discussion focuses on the lighting aspect related to breeding cabinet design.

Low voltage full spectrum LED lights were installed in each cabinet to enhance the natural light reflected into the bird room, but not to extend the normal day length. These lights are relatively inexpensive and are commonly sold for indoor plant growth. They are powered by a USB source and come with a timer to operate the lights during daylight hours. See "Sunlight & UV Light" on page 35.

The Benefits of Custom Designed Breeding Cabinets

A custom-designed breeding cabinet offers several benefits for both the birds and their caretakers.

1. Enhanced Health and Well-being

    A well-designed cabinet provides a safe and stimulating environment that promotes the health of the birds. With ample space, proper ventilation and a good environment, Gouldian Finches are more likely to thrive and exhibit natural behaviours.

2. Increased Breeding Success

    Custom cages can significantly increase Gouldian breeding success rates. Features such as the nesting areas, temperature control and ample feed options will all contribute to a better breeding environment.

3. Ease of Maintenance

    Custom cabinets are designed with accessibility and ease of maintenance in mind. Removable trays, large doors, self-filling feeders and easy-to-clean materials make daily care and cleaning more efficient.

Conclusion

By considering the specific needs of the birds and incorporating features that enhance their environment, caretakers can ensure the health and successful breeding of Gouldian finches. Whether a seasoned breeder or a beginner, a custom-designed cabinet can make a significant difference in the overall experience and success.

Colony of Red-Headed Gouldians with two new fledglings

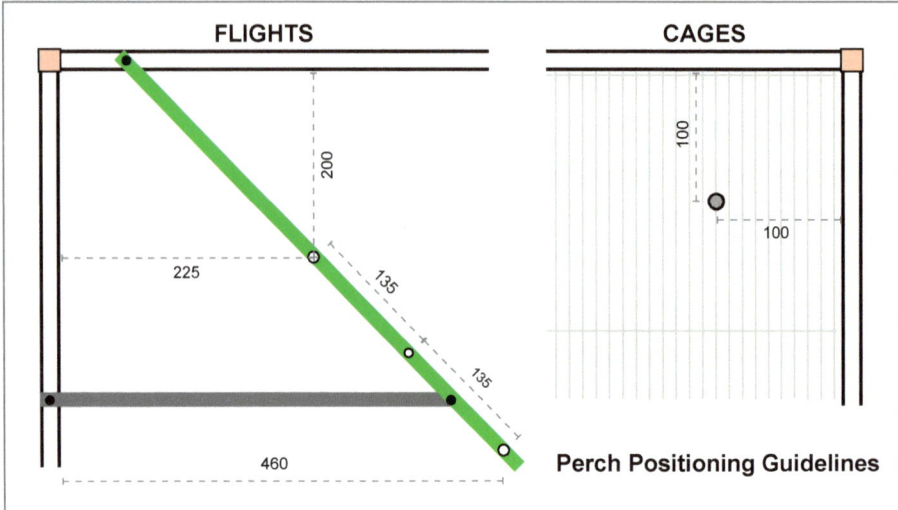

Suggested measurements for the positioning of perches (mm)

Positioning of Perches

When positioning the perches in a flight or breeding cabinet it is important to avoid placing them above any food or water.

Perches should also not be too close to the end walls of a flight, or too close to the ceiling - especially for birds in a cabinet.

> **TIP** To avoid soiling food, water or other birds always be aware of the "Drop Zone" below perches.

Bird Behaviour and Roosting

When designing an aviary, the height of perches is important in affecting bird behaviour and roosting patterns due to their sensitivity to their environment.

Gouldian finches are very "predator aware" and they have the have the instinct to prefer for elevated perches, as they offer security and protection. In the wild, higher perches allow birds to monitor their surroundings and escape quickly if needed. In an aviary or flight, providing perches at different heights can mimic this natural behaviour, increasing the birds' sense of safety and comfort.

Roosting is a critical aspect of bird life and the height of perches can significantly influence roosting behaviour. This is because birds prefer to roost in locations that offer safety and minimal disturbance, the elevated perches can provide these conditions.

This means that the highest perches should be placed where the birds are intended to roost at night. If one end of a flight provides a protected environment for night, those perches should be positioned approximately 250mm (10 inches) higher than the ones in the more exposed locations.

Minimum Measurements

Suggestions for the minimum measurements when positioning perches are shown in the "Quick Reference Table - Useful Standard Dimensions" on page 50. These will help to

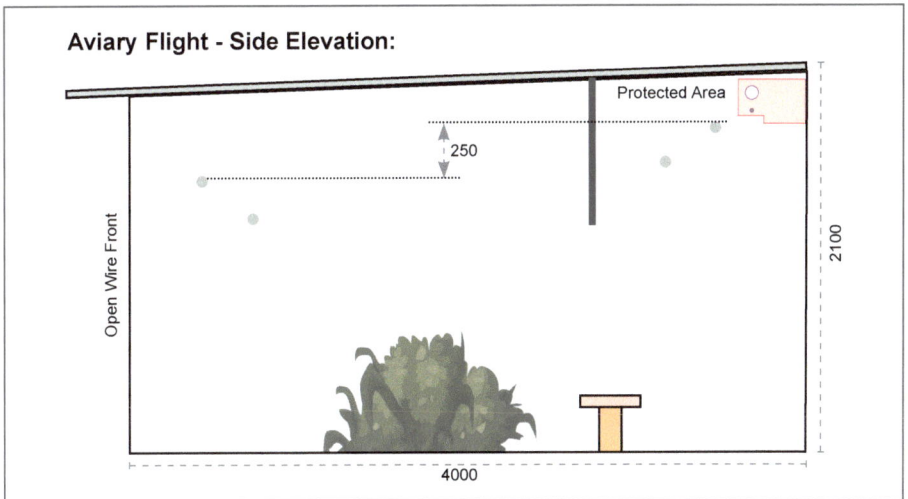

Perch positioning to encourage safe roosting behaviour

avoid droppings landing on the walls, or falling onto birds below; as well as allowing the birds to stand naturally with sufficient head room.

Quick Reference Table - Useful Standard Dimensions

	Description	Size (mm)	Size (ft, in)	Comment
Breeding Cabinets	Length	1000	3 ft, 3 in	Long enough for flying
	Height	525	1 ft, 9 in	
	Depth	450	1 ft, 6 in	
	Perch from End	100	4 in	Room for tails not to hit walls
	Perch from Roof	100	4 in	Ample room for birds to stand
	Foraging Tray	180 x 220	7 x 9 in	
	Tray Door Height	40	1½ in	Secure access to the dirt tray
Flights	Length	3000	10 ft	Long enough for flying
	Height	1900	6 ft, 4 in	Room for owner to stand
	Width	1500	5 ft	Room to catch birds when needed
	Perch from End	225	9 in	Room to avoid droppings on walls
	Perch from Roof	200	8 in	Room to avoid roof heat
	Perches Angled	45°	45°	Sets of horizontal perches
	Perch Separation	135	5½ in	
	Bird Wire	6	¼ in	Vermin wire, coloured black for visibility
	Stress Perches	100	4 in	Width, height & depth of each one
Nest Boxes	Length Overall	255	10 in	
	Length of Chamber	170	7 in	Length balance of 85 is the veranda
	Width	140	5½ in	
	Height of Chamber	165	6½ in	Deeper to protect chicks
	Height of Veranda	120	5 in	
	Chamber Opening	60	2½ in	Diameter
Other Items	Coop Cups (diam)	75	3 in	150ml capacity
	Catching Net (diam)	300	12 in	Manoeuvrable size
	Leg Rings (diam)	2.5	1/10 in	

Some useful guidelines during housing construction

NUTRITION & FEEDING

Gouldians are grass-finches, so their diet consists primarily of grass seeds, supplemented with fresh greens.

Basic Diet

Mixed Seed

A balanced diet for Gouldian finches includes a high-quality seed mix, usually sold as "Finch Mix". It is essential to provide a variety of seeds to meet their nutritional needs and the typical proportions in a suitable mix are as follows. .

Seed Variety	Typical Finch Mix A	Typical Finch Mix B	Typical Budgie Mix	Gouldian Recommended
Plain Canary	25%	25%	35%	30%
French White Millet	20%	25%	35%	30%
Red Panicum	20%	25%	15%	25%
Yellow Panicum	20%	0%	0%	0%
Japanese Millet	15%	25%	15%	15%

Ref: "Typical" results are based upon physical seed counts

Even though different products are all labelled "Finch Mix", the proportions of different seeds varies - with some excluding tonic seeds so breeders can add preferences if they wish. A simple experiment is recommended: Offer a choice of alternative mixes in different containers and measure the consumption. When this experiment was conducted in the author's own aviary the Finch Mix B was more popular than Finch Mix A and Budgie Mix.

It is also an easy task, but a little time consuming, to separate the seeds in a sample of a preprepared mix. This can be useful when evaluating a new supplier; especially since many do not disclose the proportions used. (Shirohie Millet is a specific type of Japanese Millet).

Another option is to use a "Budgerigar Mix" for the standard Gouldian seed. This has the advantage of a higher proportion of Plain Canary and French White Millet seeds than a typical "Finch Mix", which are popular with many Gouldian finches.

Austerity Seed

Gouldian finches face various challenges throughout the different seasons. Environmental conditions act as a natural selection mechanism; and when the wet season begins the increase in food resources indicates a suitable time for breeding.

To mimic this natural response, many breeders of Gouldians offer a limited variety of seeds to simulate the austere phase. Austerity feed typically includes a simpler mix of seeds low in fats and protein, which closely resembles the scarce resources available during the dry season in their natural habitat.

Components of a typical Finch Mix

Breeders commonly use a basic mix of millet and canary seed, without including green food or seeding grasses. This approach helps to simulate the natural cycles between scarcity and abundance, encouraging breeding patterns when the diet is later enriched.

Greens & Grains

Gouldians are grass-finches, so a mixture of grass-seeds, often called "Greens & Grains", will be popular – especially when promoting the breeding season. Contents of this mix includes Silk Sorghum, Signal Grass, Phalaris, Hemp seed, White Lettuce, Ryegrass seed and Cocksfoot (Kanulgras).

Greens & Grains

Nutrition & Feeding

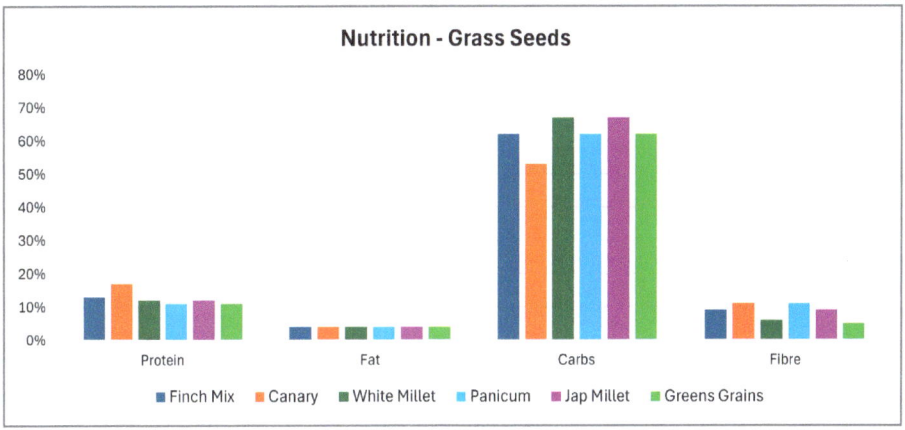

Seed Storage

To keep seed clean and fresh it is important to store it carefully. An airtight container is best for freshness and it has the advantage of protecting the seed from mice or insects. Another potential problem is webbing moths. For more information see "Webbing Moths in Seed Hoppers" on page 174.

Inspect new sources of seed for insect infestations

Supplements

Supplements can be used to ensure that Gouldian finches receive all necessary nutrients for optimal health. Some supplements should be always available, like shell grit, while others are provided as needed.

Examples of when supplements are often given include when the birds are not receiving enough natural sunlight (Vitamin D); or because particular mutations are not able to produce everything they need; or to improve the condition of weaker birds; or as a disease preventative for health; or as a boost after bad weather.

The products available and their brand names vary in different countries around the world. It is recommended to learn about your local alternatives by using Google or speaking to members of your bird club, a veterinarian, or a reputable pet shop specialising in birds.

Calcium Supplements

Calcium is an important mineral for all birds, particularly during breeding. It is vital in bone formation, eggshell production, muscle function and nerve transmission. Therefore, maintaining adequate calcium levels is necessary for breeding Gouldian

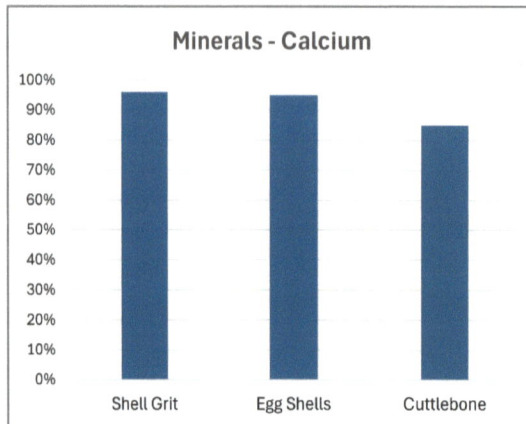

adults and their chicks.

Bird eggshells consist mainly of Calcium Carbonate, offering the necessary strength to safeguard the developing embryo. A lack of sufficient calcium can lead to thin or soft eggshells, which increases the likelihood of egg breakage and lowers the hatching success rates.

During the breeding season, Gouldian finch owners are advised to provide supplemental calcium sources. These supplements include cuttlebone, calcium blocks, oyster shell grit and calcium-rich foods such as dark leafy greens.

Diatomaceous earth is mainly used to reduce mite infestations and control webbing moths in seed. However when it is added to seed it becomes an added source of calcium at 19% of volume. (It is also 3% magnesium, 5% sodium, 2% iron & 33% silicon).

An alternative method is to incorporate calcium drops into the drinking water, or to apply calcium powder to food, such as soaked seeds or seeding grasses.

Cuttlebone

This is the dried skeleton of the cuttlefish and should be provided routinely in all flights and cabinets. It serves as a source of calcium and other minerals essential for their health. Cuttlebone should be replaced if there is any sign of mould on its' surface.

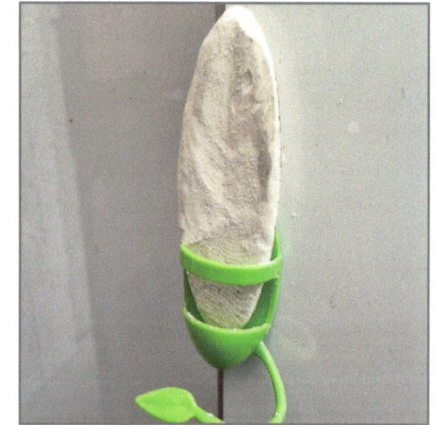

Cuttlebone in a commercial holder

Eggshells

Chicken eggshells can be an excellent source of calcium for Gouldian finches. To prepare, wash the eggshells thoroughly to remove any residue, then bake them in the oven at 180°C (350°F) for about 10 minutes. Alternatively in a microwave on "High" for 2

Baked egg-shells of chickens

minutes. Once cooled, they can be presented whole to the birds, or crush them into a fine powder and sprinkle a small amount over the seed mix or soft food. This helps ensure that the finches receive adequate calcium, which is vital for strong bones and egg production in breeding birds.

Multivitamins

These supplements are often used for birds housed in aviaries and cages due to the limited access to natural sources of vitamins, such as sunlight and various insects and plants - which all provide vitamins and minerals. Bird multivitamins typically include vitamins A, E, D3, and B-complex. Of these, vitamin D3 is particularly important; along with calcium.

Example: Inca "Ornithon", Vetafarm "Soluvite D", Vetafarm "D Nutrical Powder", Bird Health "DufoPlus", Passwell "Multi-Vite", Morning Bird "Calcium Plus" and Vetark "Nutrobal".

Another consideration is whether the product selected is in liquid or powder form. Gouldian finches de-husk their seed, so a higher concentration of powder is necessary to compensate for the wastage on those husks.

TIP If a powdered supplement is to be added to food, the dose for soft food should be doubled for dry seed without oil that will be de-husked..

Vitamin D3

This is the form of vitamin D produced by exposure to sunlight. Studies on poultry have shown that sufficient Vitamin D3 can be produced for chick growth with 11 to 45 minutes of direct sunshine per day (as reported by Naturally For Birds).

A supplement is especially important when birds have only limited access to natural sunlight or UV, which is essential for vitamin D3 production. While artificial lighting may be useful for imitating natural sunlight it is still necessary to provide vitamin D3.

	EXAMPLES: VITAMIN D3 CONCENTRATIONS IN DRINKING WATER				
	Calcivet	**Liquid Gold**	**Multi-Vite**	**Ornithon**	**Solaminovit**
Maker	Vetafarm	Passwell	Passwell	Inca	Allfarm
Category	Calcium & D3	Calcium & D3	Multi Vitamin	Multi Vitamin	Amino Acids & Multi Vitamin
Format	Liquid	Liquid	Powder (Dissolved)	Powder (Dissolved)	Liquid
Vitamin D3	25,000 IU / L	25,000 IU / L	15 mg / Kg	2.5 mg / Kg	200,000 IU / L
Vitamin D3 / L	25,000 IU	25,000 IU	600,000 IU	100,000 IU	200,000 IU
Label Dosage	5ml / 250 ml	20ml / L	1g / L	4g / L	2ml / L
Dose ml / L	20	20	1	4	2
D3 / L of Water	**500 IU**	**500 IU**	**600 IU**	**400 IU**	**400 IU**

More information: See "References Used in This Book" on page 236.

Calculations for illustration only. Always reconfirm from current information.

Most of the multivitamins also include vitamin D3. However, Cod Liver Oil is a health risk if it has become rancid.

Example: Vetark "ZolCal-F", Vetafarm "Calcivet", Bird Health "Zade", Passwell "Liquid Gold" and Wombaroo "The Good Oil"".

Most Gouldian finch owners use a general multivitamin, but it's crucial to ensure enough Vitamin D3 is included, especially for birds kept under UV-blocking materials. A typical concentration in supplements is around 500 IU of vitamin D3 per litre, with a sample analysis shown in the table:

Amino Acids

These are important for the growth, development and overall health of birds. They act as the building blocks of proteins, which contribute to muscle development, feather growth and the proper functioning of the immune system. In a caged environment, birds might not obtain all the necessary amino acids from their diet alone, particularly if their diet consists mainly of seeds. Supplementing their diet with amino acids can help ensure they receive a balanced intake, as well as aiding in reducing stress and improving digestion.

For Gouldian finches, amino acids are typically supplemented in vitamin combination products.

Combined Amino Acids and Vitamins

These supplements are high-quality, water-soluble formulations designed for poultry and other livestock. They include amino acids and vitamins A, D3, E, B1, B2, B6, B12, Nicotinic Acid, D-Pantothenic Acid, and Folic Acid.

They are used for deficiencies and also for relieving behavioural stress. Many Gouldian breeders use them as a standard dietary supplement in water or as a powder on seed. Since Gouldian finches de-husk their seed a higher dose is needed when supplementing feed rather than water.

Example: Allfarm "Solaminovit", Vetafarm "Multivet", Bird Health "Turbobooster", Naturally for Birds "Protein Boost", and Birdcare Company "Daily Essentials3".

Once again, an increased dose compensation is necessary if a powdered form is used for birds like Gouldians that de-husk their seed. For example a dose of 2.5g per kg in consumed food would be increased to 5g per kg due to the wastage of de-husking.

Shell Grit

For birds in captivity, shell grit serves as an important dietary supplement. It provides a good

Shell grit available all year

source of calcium, which is necessary for the development and maintenance of strong bones and the production of eggshells in breeding birds.

Secondly, shell grit also helps with digestion. Birds do not have teeth to grind their food, but use part of their stomach called the gizzard. Here the ingested grit mechanically breaks down food particles, improving digestion and nutrient absorption.

Additionally, shell grit can provide behavioural advantages. Pecking at the grit acts as a natural foraging activity, which keeps birds mentally stimulated and engaged. This may help to reduce stress and prevent behavioural issues such as feather plucking and aggression.

Example: Naturally for Birds "Fit Grit", Peckish "Shell Grit" and Probird "Shell Grit".

Electrolytes

These are essential minerals that dissolve in water and carry an electric charge. They are necessary for various bodily functions, including maintaining fluid balance, supporting nerve and muscle function and regulating the body's pH levels. In caged birds, electrolytes help prevent dehydration and ensure proper physiological functions, especially during periods of stress, illness, or extreme temperatures.

Example: Vetafarm "Spark", Bird Health "Quik Gel", Vetsense "BirdCare" and Nekton "Electrolyte for All Birds".

Omega 3 & 6

These supplements contain the Omega-3 & 6 essential fatty acids that are often deficient in seed mixes, together with vitamins including A, D3 & E. They are claimed to enhance breeding performance by assisting with egg production, chick growth and immunity.

Example: Wombaroo "The Good Oil", Vetsense "Avi Vital" and Aristopet "Vitamin Drops".

Iodine

Iodine is a trace element that can be deficient in the diets of aviary birds, so it is discussed more as a health issue - particularly related to feather loss in Gouldians. For more information see "Iodine Deficiency in Feather Loss" on page 170.

Apple Cider Vinegar (ACV)

The use of this product is discussed in the chapter called "Health & Wellness". See "Apple Cider Vinegar" on page 91.

Recommended Levels for Vitamin Supplements

Underdosing & Overdosing

While there are health risks if birds don't get enough of the essential vitamins and minerals, there are also risks if they get too many. These conditions are known as "hypovitaminosis" and "hypervitaminosis" respectively.

Recommended Concentrations

All aviary birds require supplements of vitamins and minerals to achieve their natural

	DEFICIENCY Underdosing	**TOXICITY** Overdosing
Vitamin A Retinol, Beta-carotene	• Low immune function • More vulnerable to infections • Skin conditions • Respiratory issues • Poor breeding results • Vision problems (Hypovitaminosis A)	• Lethargy • Reduced appetite • Impaired liver function (Hypervitaminosis A)
Vitamin B Complex Thiamine, Riboflavin, Niacin, Pantothenic Acid, Pyridoxine, Biotin, Folate & Cobalamin	• Low energy level • Feather abnormalities and poor quality • Low growth rate • Movement difficulties or seizures • Inflammation around the beak	• Lethargy • Inability to perch • Tremors or twitching • Poor coordination • Diarrhoea
Vitamin D3 Cholecalciferol	• Low calcium absorption • Fragile or deformed bones • Weak or soft-shelled eggs • Difficulty laying eggs • Decreased vitality (Hypovitaminosis D)	• Hypercalcemia (See overdosed Calcium below) (Hypervitaminosis D)
Vitamin E Tocopherols, Tocotrienols	• Poor breeding results • Muscle degeneration and stiffness • Tremors and coordination difficulty • Poor hatchability of eggs (Hypovitaminosis E)	• Impaired immune response • More vulnerable to infections • Oxidative stress and poor health (Hypervitaminosis E)
Vitamin K Phylloquinone, Menaquinone	• Excessive bleeding after injury (Vitamin K Deficiency Bleeding)	• Bleeding disorders • Excessive blood clotting (Hypervitaminosis K)
Calcium Calcium Carbonate, Calcium Citrate	• Brittle bones leading to fractures • Weak eggshells • Egg binding in hens • Poor growth in chicks • Poor breeding results (Hypocalcemia)	• Calcification of tissues • Kidney damage • Kidney stones • Disorientation and distress (Hypercalcemia)
Iodine Potassium Iodide, Sodium Iodide, Kelp	• Poor feather quality • Feather loss and balding • Lethargy • Reduced appetite • Thyroid gland enlargement (goitre) (Hypothyroidism)	• Weight loss despite adequate food • Restlessness • Reduced fertility • Poor breeding results • Thyroid gland enlargement (goitre) (Hyperthyroidism)

Vitamins and minerals are essential, but greatly excessive amounts can also cause health problems

RECOMMENDED CONCENTRATIONS			
	Maintenance	Breeding/Moulting	Ref
Vitamin A (IU)	5,000	11,000	18,19
Vitamin D3 (IU)	1,000	3,000	17,18, 19,24
Vitamin E (IU)	10	50	18,20, 21
Calcium (mg or mcL)	2,000	5,000	17
Iodine (mg or mcL)	2.5	2.5	29

Units per L or Kg. See "Vitamins & Minerals in Supplements" on page 227.

diet. Dietary shortages or excessive over-dosing can both result in health problems. For example, a lack of Vitamin D3 can result in thin shelled eggs or bone fractures. However too much Vitamin D3 can cause calcification and kidney disease.

The table shows the recommended concentrations to be supplied to Gouldians in aviary and cabinet environments.

Most recommendations for minimum concentrations of vitamins and minerals have been well published and a selection of these references are included in the "Vitamins & Minerals in Supplements" on page 227.

Recommended Level for Iodine

However, iodine can be more difficult to obtain a recommended guideline. Fortunately there is useful and relevant information in the book "Comparative Avian Nutrition" by Kirk Klasing:

He states that 1 Kg of small birds require a minimum of 0.3mg of iodine per day. Given that they drink 120ml per day, they will require that amount of 0.3mg in 120ml - or a concentration of 2.5mg/L.

Just as a confirmation, looking at this as one individual Gouldian finch weighing 15g, it will require 0.0045mg of iodine and drink 1.8ml per day. This amount of 0.0045mg/1.8ml is again the same concentration as 2.5mg/L.

It is also interesting to note that Klasing states the toxic level of iodine in small birds is 100mg for a Kg of birds - a level that is 333 times greater than the minimum of 0.3mg.

Other Sources of Vitamins & Minerals

It must be remembered that not of these important vitamins and minerals will need to be provided in a supplement. As previously mentioned, exposure to 11 to 45 minutes of direct sunlight (depending on the season) will provide all of the necessary Vitamin D3 in a natural way.

So some Vitamin A will also be provided in green leafy vegetables; D3 in sunshine; E in seeds and leaves; and Calcium in the form of shell grit and baked eggshells.

EXAMPLES: FINAL DELIVERY CONCENTRATIONS OF VITAMIN A					
	D Nutrical	Multi-Vite	Ornithon	Soluvite D	The Good Oil
Maker	Vetafarm	Passwell	Inca	Vetafarm	Wombaroo
Format	Powder	Powder	Powder	Powder	Liquid
Label Dosage	5g / Kg	1g / L	4g / L	4g / 400ml	15ml / Kg
Vit. A Ingredient	500,000 IU	1,333,333 IU	1,000,000 IU	100,000 IU	50,000 IU
Vit. A Delivered	**2,500 IU**	**1,333 IU**	**4,000 IU**	**10,000 IU**	**750 IU**

Ingredients and Final Delivered are per L or Kg.   For complete table see "Vitamins & Minerals in Supplements" on page 227

Calculations for illustration only. Always reconfirm from current information.

Final Delivery Concentrations

Some supplements address specific needs, so they are designed to be used in conjunction with other products. However, some other supplements cannot be combined with one another because the ingredients and dosage rates mean that the total amounts delivered would be too high.

It is therefore important to understand the concentrations of vitamins in different products, but the figures on the product labels cannot be compared directly. They are simply the concentration in the product itself. A product with 1,333,333 IU might seem like it will provide too much of a vitamin when compared to one with 500,000 IU, but if the first is added at a lower rate than the second then the result could be the opposite to what was implied.

The formula for calculating the concentration delivered is as follows:

Dose / Volume	X	Product Concentration	X	Days per Wk / 7	=	Concentration Delivered

For example:
5 / 1000	x	500,000 IU	x	7 / 7	=	2,500 IU

The table shows some calculation examples of these Final Delivery Concentrations. Also see "Vitamins & Minerals in Supplements" on page 227.

Undisclosed Concentrations

Some supplement products state that they contain specific vitamins and minerals, but they do not specify the concentrations. In these cases, experienced breeders of Gouldian finches generally prefer to use products where information is provided, so they can confirm that they are providing a sufficient concentration, without over-dosing.

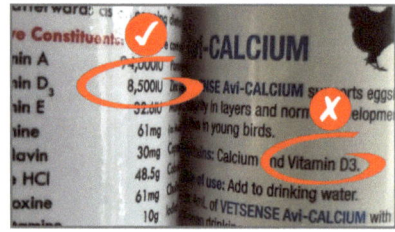

Concentrations are not always shown

Feeding Tips

Offering fresh food daily and ensuring clean water is available are fundamental practices in maintaining the health of your birds. However these are not the only areas to consider for the best nutrition of Gouldian finches . .

Live Food

Live food, such as insects, are vital component of the diet for many finches, but this is not essential for Gouldians. In fact, for Gouldian finches to eat live food they need to learn from other birds. If they do learn, it helps to mimic the natural diet of these birds in the wild, where they would consume various insects along with seeds and plants.

Gouldians can learn to eat high protein Mealworms

Live food is an excellent source of protein, which is essential for growth, muscle development, and overall health. It can be particularly beneficial during breeding seasons, as the high protein content supports egg production and the growth of chicks. Common options for live food include mealworms, small crickets, and fly larvae. If live food is fed it must be in moderation to ensure a balanced diet alongside seeds, vegetables and other nutritional elements.

For more information about setting up a mealworm colony see "Supplying Live Food" on page 177.

Dried Live Food

Dried insects are not technically "live", but they are convenient and they still offer excellent nutrition. The data for components like protein will appear higher only because the moisture has been removed from the calculation. For example, live mealworms are listed as 20% protein while dried ones are 53%. This is because live mealworms are 62% moisture, versus 5% when dried. So the protein is the same, but it's a higher percentage of a lower weight. (See "Nutrition Facts" on page 226).

The challenge with all forms of live food is to get Gouldians to eat them. When dried mealworms are offered, Gouldians need to learn to eat them form other birds in the colony. Crumbling the dried mealworms and mixing with seed is a good way to start.

Vegetables

These are a normal component of the diet for Gouldian finches as they provide vitamins, minerals, and fibre that contribute to their health. Leafy greens such as spinach, kale, bok-choy, watercress, chickweed, endive, chicory, snow pea sprouts and lettuce can be beneficial. They should be offered fresh as leaves or finely chopped to ensure the finches can easily eat them. Other vegetables like Lebanese cucumber, carrots, broccoli, corn, pumpkin and peas can also be provided, but the birds may have to observe other birds to accept these options.

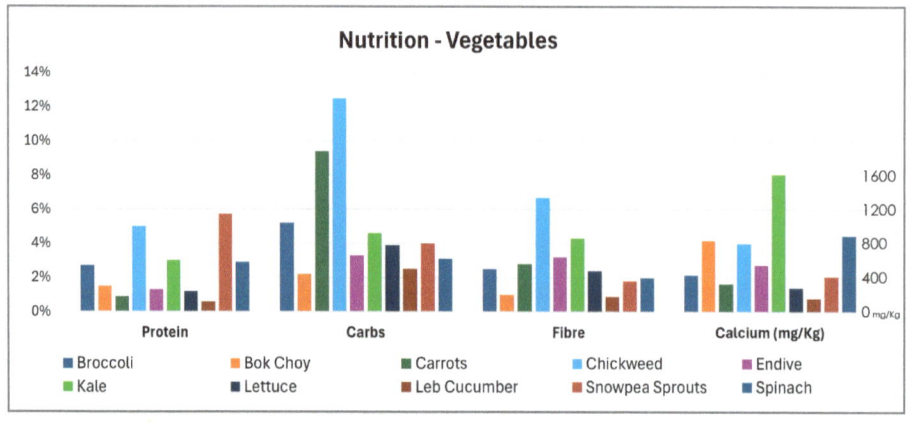

This photo of Chickweed is helpful for identification

Fresh pea sprouts are convenient and nutricious

A variety of vegetables should be a supplement to their primary diet of seeds, mimicking the natural diet. Small amounts should be offered to prevent spoilage and they should be limited to weekly in order to avoid digestive issues, diarrhoea and scouring.

Feeding spinach and kale too frequently can lead to nutritional imbalances and excessive Oxalic Acid, which may interfere with vitamin D processing and calcium absorption. This issue is less significant if the Gouldians are exposed to direct sunlight, but in cabinets these vegetables are typically provided no more than once per week.

Regularly rotating the types of vegetables provided can keep the diet interesting and nutritionally balanced for the finches. It is also important to wash all vegetables thoroughly to remove any pesticides or chemicals.

Vegetables are not provided every day

Seeding Grasses

Millet sprays and other grass-seeds are a natural and nutritious component of the diet for Gouldian finches. These grasses provide essential nutrients and support natural foraging behaviours. Providing seeding grasses allows the finches to engage with the seeds, promoting mental stimulation and physical activity. It is recommended to hang the millet sprays within their enclosure or place them in foraging trays, ensuring they are accessible while encouraging movement and exploration.

Fresh seeding grasses are very popular and healthy

Seeding grasses give particular benefits during the breeding season, as they provide a source of carbohydrates and proteins that can support egg production and the growth of chicks. To ensure a balanced diet, seeding grasses should only be provided as a supplement to the primary diet of seeds, vegetables and other nutritional elements. Regularly rotating the types of grasses offered can keep the diet varied and interesting for the finches, promoting overall health.

Foraging Trays

Introducing foraging areas into flights and cabinets aligns with the natural instinct of Gouldian finches to seek food in the wild. Foraging trays containing scattered seeds, chopped vegetables, or sprays of seeding grasses can encourage natural behaviours that stimulate Gouldians both mentally and physically. This activity may also help prevent boredom and reduce stress in the birds.

Foraging trays encourage natural behaviour

If the flight area is large enough, foraging trays should be distributed in various locations to encourage exploration and movement. During breeding seasons, foraging opportunities provide exercise through physical activity and nutrient-rich foods, which support the energy needs of egg production and chick rearing.

Breeding cabinets equipped with foraging trays improve the lives of Gouldian finches

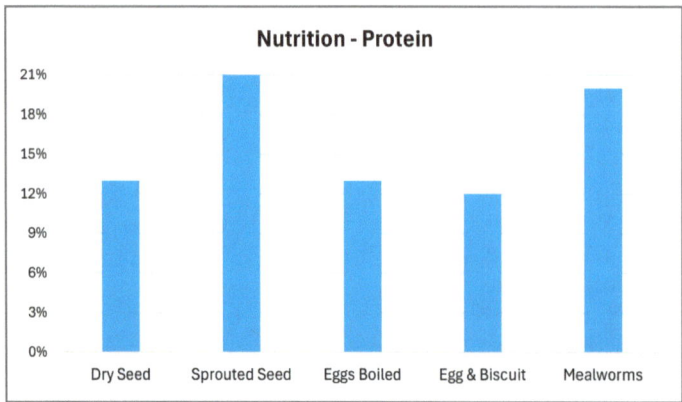

by bringing more of their normal behaviour into captivity, resulting in a more natural and fulfilling environment. Additionally, it is necessary to regularly rotate the contents of foraging trays to keep the finches engaged and interested.

Boiled Eggs

Eggs can provide an excellent source of high-quality protein, which is essential for the growth and maintenance of feathers, muscles and tissues in birds. The protein content supports the development of chicks and juveniles, but they are not essential for feeding to Gouldians - even during the breeding season.

To prepare eggs for Gouldians, first boil them until they are hard. Then allow the eggs to cool completely before mashing them with the shell still on. Once offered to the birds, any uneaten egg must be removed within 3 hours to prevent spoilage, depending on average temperatures.

Egg & Biscuit Mix

Like boiled eggs, Egg and Biscuit Mix is used by some breeders of Gouldians, but it is not essential if the other diet is sufficient. Egg & Biscuit is widely recognised as an excellent source of dietary proteins, fats, vitamins and minerals, which contribute to the overall health and well-being of birds.

The biscuits in the Egg and Biscuit Mix supply carbohydrates and fats. Carbohydrates provide energy, supporting birds' activity levels. Fats are essential for healthy skin and feathers and act as insulation and as an energy reserve.

Commercial Egg & Biscuit products are available for convenience. Alternatively, preparing the egg and biscuit mix is straightforward, requiring 1 hard-boiled egg and 2 tablespoons of biscuit crumbs, crushed wholemeal biscuits, or crumbled day-old bread.

1. Boil the egg until hard-boiled, then let it cool completely.
2. Mash the unpeeled egg with a fork until smooth.
3. Mix biscuit crumbs, crushed biscuits or crumbled bread into the mashed egg.
4. Add calcium or supplements.
5. Serve to birds in a clean feeding dish.

It is essential to ensure that the mixture is freshly prepared and not left out for extended periods, as it can spoil rapidly. Three hours is a good guideline.

Example: Passwell "Egg & Biscuit", Sheps "Egg & Biscuit" and Avi One "Egg & Biscuit".

Blended Calcium Grit Products

Although it is possible to purchase shell grit separately, a better alternative is to opt for a blended product. These products typically contain a combination of ground calcium grit from various sources, sterilized eggshell, cuttlebone, as well as essential minerals and vitamins.

These blended products typically include Calcium Carbonate, Sodium, Iron, Cobalt, Copper, Zinc, Manganese, Iodine, Molybdenum, Selenium, Vitamins A, D3, E, B Complex, K, Pantothenic Acid, Folic Acid, Niacin, Choline Bitartrate, probiotic Protexin, activated charcoal and Diatomaceous Earth.

Example of a blended grit product

Example: Naturally for Birds "Fit Grit".

Soaked & Sprouted Seed
Soaked Seed

These seeds are dry seeds that have been immersed in water for 12 to 24 hours. This process begins germination, making the seeds more digestible, easier for parents to feed to chicks and starting to increase their nutrient content, including vitamins, minerals, and amino acids.

TIP Soaked Seed is quick and easy to prepare and offers improved digestibility and easier feeding for chicks. However, Sprouted Seed offers increased protein and nutrients.

To soak the seeds they are placed in a container and covered them with fresh clean water. The water is then changed every 6 to 8 hours to prevent bacterial growth and then thoroughly rinsed with fresh water again before offering them to the birds.

Sprouted seeds give better nutrition than soaked seeds, but the preparation time is longer.

Recommended Procedure for Soaked Seed

The following steps describe how to soak seed safely . .

1. Select the Seeds: Choose fresh high-quality seeds, such as millet, canary seed, or your normal Finch mix.
2. Rinse: Use cold running water to thoroughly rinse the seeds and remove any dust or contaminants.
3. Soak: Using clean fresh water, soak the seeds for 24 hours with a water disinfectant added (chlorhexidine base is recommended, like Vetafarm "Aviclens", Passwell

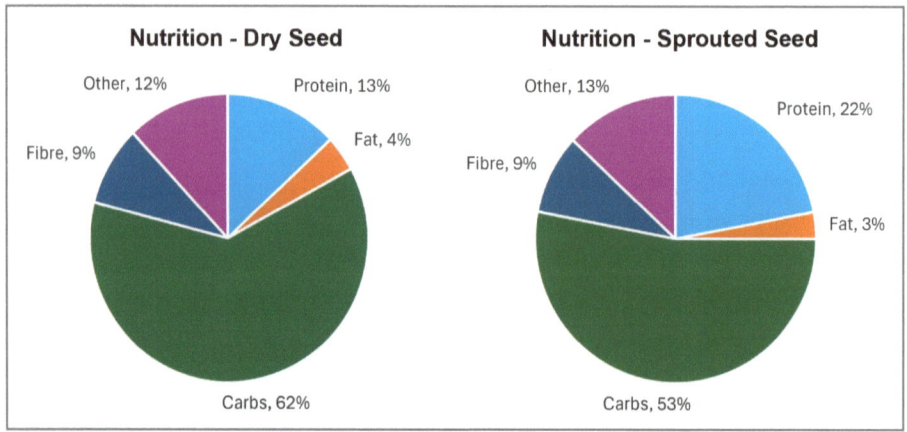
Protein is significantly increased when seed is sprouted

"Multi-Clens", or Ranvet "Virkon S").

4. **Rinse:** On day 2, drain, rinse and drain the seeds. Leave them in the sieve for adequate drainage and air circulation to prevent mould growth.
5. **Smell Test:** The soaked seeds should smell sweet to be fed to the birds. If they smell mouldy or bad in some way, discard them.
6. **Serve:** On day 3, the soaked seeds can be served to the birds
7. **Storage:** Any unused soaked seeds can be stored in the refrigerator for up to 3 days, but always check for freshness and spoilage before feeding them to the birds.

Soaked seed during preparation

Sprouted Seed

This type of seed has numerous benefits for Gouldian finches. When seeds have sprouted, or chitted, they are more nutritious than dry seeds because the growing process increases their nutrient content and makes them easier to digest. They have increased protein and are rich in vitamins, minerals, and enzymes. They can therefore enhance the overall health and vitality of the finches. Additionally, sprouted seeds provide a source of hydration, which is especially beneficial in dry conditions. Offering sprouted seeds regularly can improve the birds' immune system, promote healthy plumage, and support their growth and reproductive health. It is especially of benefit when Gouldians are feeding chicks and fledglings.

Nutrition & Feeding

Recommended Procedures for Sprouted Seed

Many ways have evolved for safely sprouting seed, so the following steps describe two different popular methods. Procedure No 1 is a classic method . .

1. Select the Seeds: Choose fresh high-quality seeds suitable for sprouting, such as millet, canary seed, or your normal Finch mix.
2. Rinse and Soak: Rinse the seeds thoroughly under cold running water to remove any dust or contaminants. Then, soak the seeds in clean, fresh water for 8 to 12 hours overnight.
3. Drain and Rinse: After soaking, drain the seeds and rinse them again to ensure they are clean.
4. Drain Again: Leave the drained seeds in a shallow container with holes that allow for adequate drainage and air circulation to prevent mould growth. A sieve is a good option for this.
5. Keep Moist: Use a spray bottle to spritz the seeds with clean water a few times

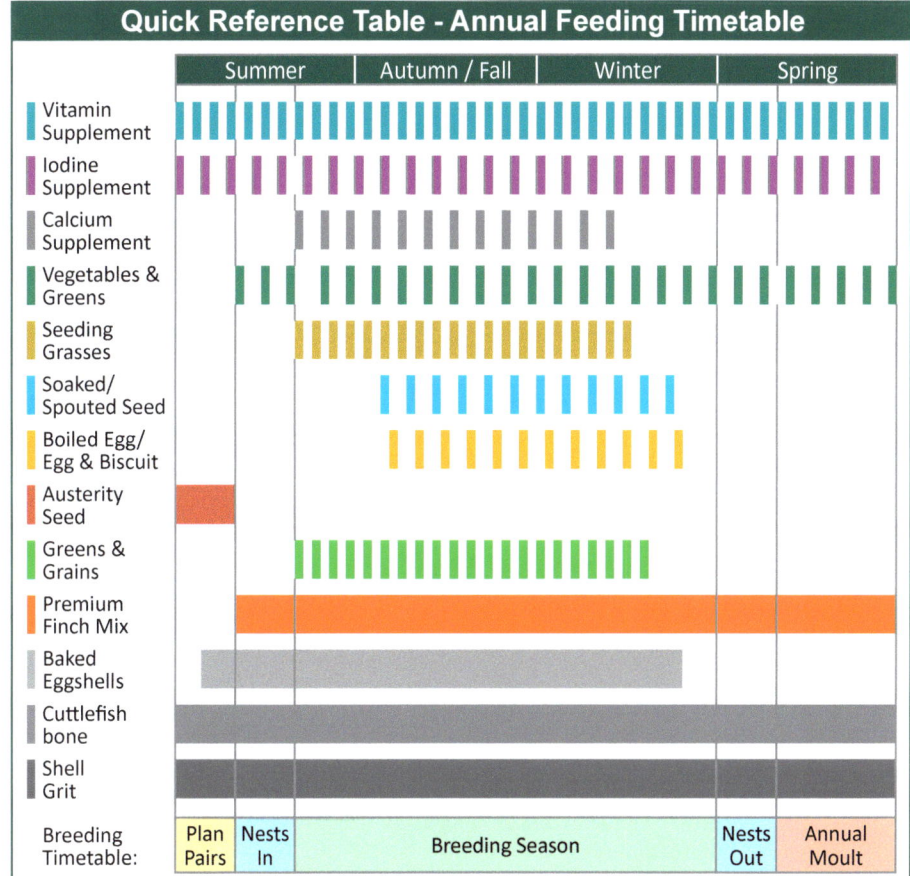

The annual feeding timetable for Gouldian finches is based around the Breeding Timetable

Gouldian Finches - Care, Breeding & Genetics

a day, keeping them moist but not waterlogged. Using the sieve allow excess water to drain away to avoid standing water.

6. Monitor, Rinse and Drain: Continue rinsing and draining the seeds 2 times a day. Depending on the temperature, you should see tiny sprouts emerging. This is called "chitted" and the seeds are ready to feed when the sprouts are about 3mm (1/8") long.

7. Final Rinse: Stand the seeds in water treated with "Multi Clens", "Aviclens", or "Virkon S", or chlorine for 30 minutes, then rinse them thoroughly under fresh running water.

8. Smell Test: The sprouted seeds should smell sweet to be fed to the birds. If they smell mouldy or bad in some way, discard them.

9. Storage: Any unused sprouted seeds can be stored in the refrigerator for up to 3 days, but always check for freshness and spoilage before feeding them to the birds.

Sprouted seed that has just chitted and is ready to feed

Many breeders of Gouldian finches also use what is seen as a more time efficient procedure to safely sprout seed. Procedure No 2 . .

1. Select Container: Use a clear container with a lid that has small drilled drainage holes.

2. Select the Seeds: Choose fresh high-quality seeds suitable for sprouting, such as millet, canary seed, or your normal Finch Mix.

3. Prepare Disinfectant Solution using "Multi Clens", "Aviclens", or "Virkon S". For example, a 0.5% solution of Ranvet "Virkon S" is a 5g teaspoon in 1 litre of water.

4. Rinse and Soak: Rinse the seeds thoroughly under cold running water to remove any dust or contaminants. Then, soak the seeds in Disinfectant Solution for 2 to 4 hours (maximum).

5. Drain: After soaking, turn the container over and drain the seeds through the holes in the lid.

6. Stand & Stir: Leave the container in a naturally lit area, but not in direct sunshine. Stir twice a day for even seed growth.

7. Chitted: After 24 hours the seeds will be just starting to sprout by 1mm. This is called "chitted" and they are now ready to feed to the birds. (This may take longer in colder weather).

8. Leave Dry; The seeds do not need to be rinsed again.

9. Smell Test: The sprouted seeds should smell sweet and "nutty" to be fed to the birds. If they smell mouldy or bad in some way, discard them.
10. Storage: Any unused sprouted seeds can be stored in the freezer for 3 months, but always check for freshness and spoilage after thawing and before feeding to the birds.

Similar to Egg & Biscuit Mixes, it is common for Gouldian breeders to add calcium or supplements to soaked or sprouted seed.

Special Diets for Special Birds

It is often believed that certain mutations of Gouldian finches necessitate specific diets or supplements. These views may be grounded in facts or speculation. However, given that many breeders engage in varying degrees of in-breeding to increase the population of these mutations, the use of supplements is likely well-founded.

Supplements for the Blue-Backed Mutation

Regarding the blue family of mutation (blue, pastel blue, silver, etc), some breeders have suggested that these birds require special diets, while other breeders do not differentiate their care from other Gouldian finches.

Those that do care differently for the blue Gouldian finch mutation have often noted its' susceptibility to certain dietary vitamin deficiencies, which can impact overall health. These genetic predispositions include vitamins A, D, E and B. Whether this is because they have increased metabolic demands, or they have altered nutrient absorption and utilisation, it is common for breeders of the blue mutations to include specialized supplements targeting vitamin needs.

Pastel blue mutation

Supplement combinations such as Allfarm "Solaminovit" and "Solvita AD3 High E"; Vetafarm "Soluvite D" and "D Nutrical"; or Bird Health "DufoPlus", "Ioford" and "Zade", are all commonly used. However, these are concentrated forms of vitamins and they are not recommended for constant daily use due to the risk of liver damage from overdosing.

Containers for Feeding Seed

The choice of containers for seed is one between open dishes and feed hoppers.

Open Dishes

Open dishes replicate the natural foraging conditions, but they need to be refilled often and the seed husks need to be blown away from the surface so that the birds can read the seeds.

The containers also need to be cleaned regularly to maintain hygiene and protect the health of the birds. For those with busier lifestyles open dishes are therefore not popular.

A hopper design liked by the owner compared to one that is more popular with the birds

Feed Hoppers

Seed hoppers offer greater convenience since they are designed to refill themselves, however they have different challenges to consider. The seed in the hopper must be free flowing in order for it operate as intended. Seed clogging can occur if oily supplements have been added to the seeds, or if there are moths living in the seed and creating "webbing".

To address clogging in seed hoppers it is recommended to use powdered supplements, rather than liquid ones. The dose will be higher since the powder will be on the husk that the birds do not actually eat. (The use of "Solaminovit" is as an example).

For the challenge of moths, a bay leaf can be placed in the lid of the hopper as a natural insect repellent. Another option is garlic powder mixed through the seed.

For those people using breeding cabinets a hopper will be the better choice because they need to be refilled less often and that refilling is done from outside the cage, thus causing less disturbance to the birds. Some seed might also be occasionally placed in a foraging tray, but the reliable feed source is the automatic hopper.

The Choice of Seed Hopper Design

Trials with different seed hopper designs have shown that Gouldian finches are very "Predator-Aware". Given the choice, they will select an open dish and the open tray of a hopper, rather than a situation where they have to go inside a hopper to access the seed.

Options for Fresh Water Supply

Ensuring they have access to clean, fresh water at all times is of course vital for the health of birds in flights or breeding cabinets. As with feed, it is important to have a fail-safe system for the supply of water.

Water Bowls

Traditional water bowls or dishes are simple to use but they require regular refilling and cleaning to prevent contamination. For these containers it is best to choose bowls made of stainless steel or ceramic for ease of cleaning.

 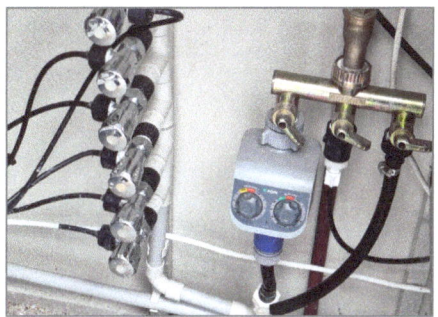

Automatic refill for water & bathing *Irrigation timers to control the watering system*

Automatic Waterers

Systems using irrigation controllers can provide a regular refilling of water with minimal maintenance. They can also be designed to include an overflow drain so that the system is flushed clean regularly. Automatic systems are particularly useful in larger aviaries or for those with busy schedules.

Water Bottles

Hanging water bottles, similar to those used for small mammals, can offer a hygienic water source and reduce the risk of spillage and contamination. They operate with a ball valve to release a drop of water when activated by the bird's beak.

Water Valves

Gravity fed Edstrom water valves also release a drop of water activated by the bird's beak. These can be gravity-fed with micro-irrigation pipes and connected to a tank that is refilled by using an automatic float valve.

Water Fountains

Some bird caretakers have used a garden re circulating fountain in an aviary or flight. These run continuously with a filter for a constant flow of clean water. They not only provide hydration, but the water movement also stimulates the birds' natural drinking behaviour.

Feeding Stations in an Aviary or Flight

A well-designed feeding station can be used to combine the various elements necessary for feed and supplements given to Gouldian finches.

Components Included

The only limitation on the design is the owner's imagination, but it is common for a feeding station to include the following:

- Seed hopper
- Coop cups
- Vegetable tray
- Nesting material
- Fail-safe seed hopper alternative
- Fail-safe water alternative
- Integrated doors for when hoppers, cups or trays are removed for cleaning.

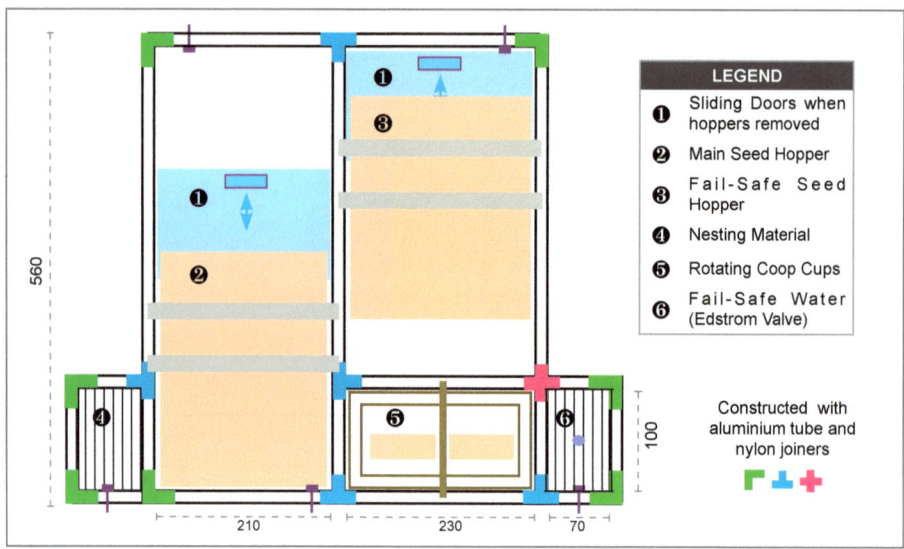
A design example for a feeding station in a flight

Positioning

The feeding station should be placed to allow birds to feed without disturbance and provide easy access to food. Note that any components positioned below another are smaller to avoid soiling from above.

Dimensions

The diagram of a feeding station designed by the author for a flight shows the typical dimensions as a guide.

Feeding station in a flight

Nutrition Data

More nutrient information for grass seeds, various protein sources, minerals and vegetables is contained the table called "Nutrition Facts" on page 226.

Standard Care Plan

It is easy to become influenced and confused by competing advice and advertising for different commercial products. It is therefore recommended to first consider the needs of your birds; then select a combination of foods and supplements to meet those needs. A form for doing such a plan is included in this book - see "Standard Care Plan for Gouldians" on page 218 - and the "Example of a completed Standard Care Plan" on page 93.

HEALTH & WELLNESS

Keeping Gouldian finches healthy ensures their well-being and longevity. Healthy finches are more active, display natural behaviours, and thrive mentally and physically. They also engage better in social interactions and breeding.

Clean Environment

Maintaining a clean environment is important for Gouldian finches as it reduces the risk of infections and diseases. Cleanliness helps control the spread of bacteria, fungi, and parasites. Regular cleaning of cages, perches, and feeding dishes ensures that the birds are not exposed to pathogens that can cause respiratory issues or other illnesses. A hygienic environment also provides a safer habitat, which encourages natural behaviours and minimizes stress for the birds.

The cleaning agents should be safe for birds to prevent any adverse effects. Clean water with mild detergent and disinfectant can be used for this purpose.

Regular maintenance of breeding cabinets every few weeks includes washing and replacing the contents of the removable dirt tray, as well as cleaning the feed and water containers.

Dry Environment

A wet floor, or high moisture levels, in an aviary or cabinet can create conditions that are a breeding ground for bacteria, fungi and parasites. Birds have sensitive respiratory systems and extended exposure to damp environments can lead to respiratory infections, such as fungal diseases that may lead to death.

In addition, wet environments encourage mould growth, thereby increasing health risks for Gouldians. By maintaining a dry environment these risks can be reduced by being pathogen-free.

Annual Cleaning

At least once a year, typically during the non-breeding season, a more thorough "deep-cleaning" is conducted. The birds are relocated, and the perches, dishes and other accessories are removed and cleaned separately to ensure comprehensive cleaning. The flight or cabinet will also be cleaned by removing all debris and using a scrubbing brush and mild detergent on all surfaces, with special attention to corners and crevices where dirt and bacteria can accumulate.

Annual Deep Cleaning	
①	Cabinets: Birds have been moved to flights -OR- Flights: Most birds have been paired in cabinets
②	Dishes & accessories removed & cleaned
③	Perches & surfaces cleaned with brush and detergent
④	Rinsed with water using pressure hose
⑤	Everything left to thoroughly air dry
⑥	Spray with ACV and leave to stand for >15 minutes
⑦	Rinse and air dry
⑧	Spray with bird-safe insecticide (eg Coopex)
⑨	Replace all accessories ready for the birds

After scrubbing, rinse the cabinet or flight thoroughly with clean water to eliminate any soap residues. Then, use a bird-safe disinfectant to eradicate any remaining bacteria or fungi. A recommended option is to use a spray bottle with a solution of Apple Cider Vinegar at a 15% concentration in water. Allow the disinfectant to sit for 15 minutes before rinsing again with clean water and ensure the area is completely air-dried before returning the birds.

All of the removed accessories are cleaned in a similar way with mild detergent, rinsing, disinfecting and rinsing thoroughly again. Then ensure they are completely dry to avoid mould development before putting them back in place.

In a practical context, it is common that breeding cabinets undergo a thorough annual cleaning at the conclusion of the breeding season. This occurs after the birds have been transferred to the flights and separated by gender into cocks and hens.

Common Health Issues

Gouldian finches can be susceptible to several health issues, including respiratory infections and mites. Recognizing and addressing these problems early can prevent severe consequences.

Respiratory Infections

Respiratory infections in Gouldian finches can arise from bacteria, fungi, viruses or parasites. Symptoms include laboured breathing, nasal discharge, and lethargy.

Early identification of respiratory infections in finches is important for effective treatment. Typical symptoms include:

- Laboured breathing or wheezing

- Nasal discharge or crusting
- Coughing or sneezing
- Swollen eyes or conjunctivitis
- Loss of appetite or weight loss
- Lethargy and reduced activity
- Tail bobbing (indicative of respiratory distress)

Bacterial Infections

Bacterial infections may cause enteritis, an inflammation of the small intestine that can then result in diarrhoea, lethargy and weight loss. Infections in the sinuses can cause visibly swollen areas on the head of the bird. Keeping a proper ventilated and clean environment together with a nutrient-rich diet can help prevent these infections.

Mycoplasma gallisepticum is a well-known bacterium that causes respiratory infections and deaths in finches.

→ Treatment again involves isolation in a hospital cage to prevent spread of disease while keeping the bird warm and stress-free. Medications are antibiotics like Inca "Sulfa 3", Vetafarm "Triple C", Aristopet "Avicycline C", MedPet "Avivet", Doxycycline and Enrofloxacin. Electrolyte solutions, like "Spark" or "Quik Gel", also aid with hydration.

Treatment Guide	
BACTERIAL INFECTION	
Antibiotic in drinking water	
Examples:	Sulfa 3, Triple C
Duration:	5-7 days

Viral Infections

Viral infections can affect Gouldian finches and be highly contagious. These include avian influenza and Newcastle disease, both of which can lead to severe respiratory distress and other clinical symptoms.

→ There are no effective treatments available for viral infections, so the only courses of action are supportive treatments like warmth and electrolytes.

Fungal Infections

Fungal or thrush infections can develop in humid conditions and may lead to respiratory problems and skin infections. Aspergillosis is a respiratory infection in birds, including finches, caused by the fungus Aspergillus. This fungus flourishes in damp, mouldy environments and can lead to chronic respiratory conditions.

Maintaining a dry and clean habitat for the finches, along with regular disinfection, helps prevent fungal growth.

Differential diagnosis is assisted because fungal infections have an unpleasant smell.

→ Treatment also involves isolation in a hospital cage and medications are anti-fungals like Nystatin, in products like or BirdPal "Fungistat" or MedPet "Medistatin". There are also natural options like Morning Bird "Revive", which is 100% Pau D'Arco bark powder.

Treatment Guide	
FUNGAL INFECTION	
Antifungal mixed in seed	
Examples:	Fungistat, Medistatin
Duration:	5-7 days
Antifungal sprinkled on seed	
Examples:	Revive
Duration:	3 days

Mite Infestations

Mite infestations, especially air sac mites (Sternostoma tracheacolum), pose a significant threat to Gouldian finches. These tiny parasites can invade the bird's respiratory system, resulting in symptoms such as coughing, wheezing and general distress. Regular cleaning and the routine use of anti-mite treatments are essential measures to reduce the presence of these pests.

→ Treatment can be for the infected bird alone, but generally it is applied to the whole flock at the same time. There are avian-safe anti-mite medications of Ivermectin or Moxidectin that can be added to the single water source during the treatment period of 24 hours. These products include Vetafarm "Scatt", Bird Health "S76" (Ivermectin and homeopathics) or Vetafarm "Moxivet Plus". A follow-up treatment is necessary 2 weeks later to ensure complete eradication of the mites.

Treatment Guide	
AIR SAC MITES	
Anti-mites mixed in drinking water	
Examples:	Scatt, S76, Moxivet
Duration:	24 hrs (repeat 2 wks)
Anti-mites individual spot-on	
Examples:	Scatt, S76, Moxivet
Duration:	1 drop (repeat 2 wks)

→ The selected treatment can be added to all available water.

→ Or it can used as a Spot-on applied to the skin behind the head or under the wing, each day for 2 days, repeated two weeks later to break the mite life cycle.

→ A prevention involves dusting cabinets, nests and floor trays with Diatomaceous Earth. See page 175.

→ A bird-safe insecticide like Coopex (pyrethrin) is also useful for controlling mites, lice, and insects before infestation occurs.

Feather Loss

Feather plucking, where birds excessively remove their feathers, causing bald patches, is a behaviour that can develop from stress. Feather loss can also be caused by the normal annual moult, a poor diet, vitamin deficiencies, parasitic infections and genetic factors. For more information see "Vitamin D3 Deficiency" on page 169 and "Iodine Deficiency in Feather Loss" on page 170.

→ Treatment consists of providing a balanced diet, Vitamin D3 supplement, Iodine supplement, medicated shell grit, full spectrum lighting, no overcrowding and stress observations. Then the feathers will regrow at the next annual moult.

Treatment Guide	
FEATHER LOSS	
Supplements in drinking water	
Examples:	Vitamin D3 & Iodine
Duration:	Ongoing

Coccidiosis

Coccidiosis is a parasitic disease that affects the intestinal tract of birds. It is caused by Eimeria protozoa, with symptoms including diarrhoea (which may be bloody), lethargy, weight loss, and reduced appetite. Infected Gouldian finches often display a fluffed-up appearance and may isolate themselves from the flock. The disease spreads quickly through contaminated droppings, so maintaining good hygiene is essential.

→ Treatment includes anti-coccidial drugs like Elanco "Baycox" and better environmental cleanliness. An example medication is Vetafarm "Coccivet", Vetsense "Cocciprol", Bird Pals "Endocox", or Bird Health "Carlox". Probiotics and a balanced diet aid recovery and strengthen the bird's immune system.

Treatment Guide	
COCCIDIOSIS	
Mixed in drinking water	
Examples:	Baycox, Coccivet
Duration:	1 Day, or 5-7 Days

Giardia

Giardia is another parasitic disease that lives in water and spreads easily to other birds through drinking and bathing. Symptoms including diarrhoea, lethargy, weight loss, excessive preening, scratching and feather plucking.

→ Treatment includes antiparasitic like Ronidazole, in products like Vetafarm "Ronivet-S" or Harkers "Harkanker" - as well as sterilising all baths and drinkers.

Treatment Guide	
GIARDIA	
Mixed in drinking water	
Examples:	Ronivet-S
Duration:	7 Days

Fractures & Trauma

Fractures and trauma are significant health issues for Gouldian finches, frequently arising from accidents, conflicts, or improper handling. Indicators of fractures include limping, an inability to perch, swelling, and visible deformities. Trauma may present as shock, lethargy, and loss of coordination.

→ Treatment is to provide immediate veterinary care for accurately diagnosing and setting fractures, often necessitating the use of splints. With consideration of the fragile size of Gouldian finches it may also be more appropriate to euthanase the bird if they are suffering? To prevent trauma, it is important to maintain a safe environment free from sharp edges and other hazards; and to handle the birds gently and infrequently to minimise stress and the risk of injury.

Eye Infections

Infections are often observed as one eye closed or by a discharge around the affected eye. Flying and landing for the affected bird will be more difficult with one eye closed and there may be frequent wiping against the perch.

→ Treatment is to place the finch in a hospital cage and treat the affected eye with a smear of antibiotic eye ointment twice a day for 5 days. Products sold for human use, such as a chloramphenicol ointment such as Aspen Pharma "Chlorsig", give good results.

Treatment Guide	
EYE INFECTION	
Ointment smeared on closed eye	
Examples:	Chlorsig
Duration:	2x/day, up to 5 Days

Nutritional Deficiencies

Deficiencies, due to a lack of essential vitamins and minerals, can cause weak bones, feather loss, poor feather condition and reproductive problems.

→ Treatment is to provide a varied diet, supplemented with vitamins, ensuring that the birds receive all necessary nutrients. The options include fortified seed mixes and supplements added to drinking water or feed.

Treatment Guide	
VITAMIN SUPPLEMENT	
Fortified seed or mixed in water	
Examples:	Solaminovit, Multivet
Duration:	Ongoing

Inadequate Natural Light

Vitamin D deficiency can be a significant problem related to housing for Gouldian finches. Adequate sunlight exposure is crucial as it allows the birds to synthesize vitamin D, which is necessary for calcium absorption and bone health. Birds kept indoors may be particularly susceptible if they lack access to UV light that acts like natural sunlight. Additionally, vitamin D deficiency may lead to symptoms such as lethargy, poor growth in juveniles and brittle feathers that break easily.

→ Treatment includes adding an artificial source of UV light, commonly known as "Growth Lights" for plants and incorporating a Vitamin D supplement into the birds' diet. For more information see "Vitamin D3 Deficiency" on page 169, in the chapter called "Common Challenges & Solutions".

Treatment Guide	
VITAMIN D DEFICIENCY	
Supplement mixed in water	
Examples:	D Nutrical, Solvita AD3
Duration:	Ongoing

Excessive Melanin

Melanism is a condition where there is excessive production of melanin, which is the pigment responsible for black and brown colours in feathers, skin and eyes. In Gouldian finches, melanism can appear as an abnormal darkening of their typically bright and varied plumage. This condition can be caused by various factors, including genetics, environmental influences and nutritional deficiencies. Reasons for Vitamin D deficiency in Gouldian finches include:

- UV Exposure: In the wild, Gouldian finches receive sunlight, enabling them to produce vitamin D naturally. In captivity, they may not have the same access to UV light, particularly if kept indoors without proper lighting. This can result in Vitamin D deficiency.
- Poor Diet: Gouldian finches need a balanced diet. Without Vitamin D-rich foods like fortified seeds or supplements, they can develop a deficiency, especially if kept indoors with limited sunlight.

This bird has Melanism after not having enough access to UV light

- Health Issues: Some health conditions can impact the absorption or metabolism of Vitamin D, leading to deficiency. For instance, liver or kidney diseases that affect the synthesis of Vitamin D in the body.

The most significant contributing factor has been identified as a lack of Vitamin D.

→ Preventing Vitamin D deficiency and melanism in Gouldian finches involves providing access to natural sunlight, or artificial UV lighting. Ensuring a diet that includes Vitamin D-rich foods or supplements can also help prevent deficiency.

Treatment Guide	
MELANISM	
Sunlight & supplement in water	
Examples:	D Nutrical, Solvita AD3
Duration:	Ongoing

Vitiligo

This condition involves excessive depigmentation due to a loss of melanin producing cells (melanocytes). The condition is not fully understood, but is believed to be autoimmune with a genetic predisposition. Gouldians with vitiligo can be identified by white depigmented feathers. Unfortunately there is no current treatment, other than removal from breeding.

Egg Binding

This is a condition in which an egg becomes lodged in the hen's reproductive tract and cannot be laid naturally. This condition is serious and potentially life-threatening, requiring immediate attention.

The causes of egg binding include:

- Poor nutrition, particularly calcium deficiency.
- Age, as both younger and older hens are more susceptible to egg binding.
- Health issues such as obesity, infections, or genetic factors.
- Environmental factors, including inadequate lighting, temperature fluctuations and lack of exercise.
- Egg size and shape, specifically abnormally large or misshapen eggs.

Signs of egg binding include straining to lay an egg without success, visible swelling in the abdomen or near the vent area, lethargy or reduced activity, laboured or heavy breathing, fluffed feathers, and decreased interest in food or eating less.

→ Treatment options can include creating a warm and humid environment to help relax the muscles, administering calcium supplements to strengthen the muscles, applying olive oil around the bird's vent, or manual extraction by a veterinarian.

Two liquid calcium drops can be applied to the side of the beak every second day to accelerate calcium absorption (eg Vetafarm "Calcivet", VetArk "ZolCal D" or Bird Health "Ioford"). In severe cases the hen may need to be rested from the breeding program for a month.

Treatment Guide	
EGG BINDING	
Liquid calcium to the beak	
Examples:	Calcivet, ZolCal D
Duration:	2 alternate days
Calcium in diet	
Examples:	Shell grit, eggshell, etc
Duration:	Ongoing

→ Preventing egg binding requires a balanced diet

that is high in calcium and other essential nutrients, along with opportunities for regular exercise, appropriate daily lighting cycles, and reduced stress levels.

Abnormal Head Movements

Abnormal head movements in Gouldian finches are actions that differ from their usual behaviours. These can appear as tilting and angling of the head to one side, sometimes causing the bird to fall over.

Several factors can lead to abnormal head movements in Gouldian finches, resulting in Twirling due to physical conditions, or Star-gazing due to environmental influences.

The most serious condition observed is Twirling, predominantly affecting Gouldian colour mutations that were bred from restricted genetic strains. Some of these birds exhibit neurological dysfunctions, leading to the development of Twirling in adulthood. Additional possible causes may include trauma and infections.

An example of Twirling in a Gouldian colour mutation

Star-gazing is more likely to be related to the enclosure's environment and may also become a habit. As these abnormal behaviours are a possible result of genetics, disease or environment, they are discussed more fully in the chapter about "Common Challenges". See "Abnormal Head Movements" on page 172.

Going Light

Various health conditions can affect Gouldian finches, with "going light" often mentioned. However, this term describes a condition rather than a disease itself.

"Going light" refers to a bird quickly losing weight despite eating. This symptom can indicate various health problems, and Gouldian finches are particularly susceptible if their needs are not met.

Causes

1. Parasitic Infections - A significant cause of weight loss in Gouldian finches is Coccidiosis and internal parasites, such as worms. These parasites deplete essential nutrients from the bird's body, leading to rapid reduction in weight.
2. Nutritional Deficiencies - Insufficient nutrition can also be a factor in birds experiencing significant weight loss. These birds need a balanced diet to meet their metabolic requirements.
3. Bacterial and Viral Infections - Bacterial or viral infections can weaken Gouldian finches' immune systems, making them prone to going light. Respiratory infections especially harm their health, reducing appetite and energy.

4. Environmental Stress - Overcrowded cages, poor hygiene and inadequate ventilation can cause going light in Gouldian finches. They thrive in a clean, stable environment with ample space to fly and interact.

Symptoms

Recognizing the symptoms of going light is crucial for early intervention. Common signs include:

- Noticeable weight loss and a prominent keel bone
- Lethargy and reduced activity levels
- Feather plucking or poor feather condition
- Decreased appetite and reluctance to eat
- Diarrhoea or abnormal droppings

Treatment

An avian veterinarian can perform examinations to diagnose the underlying cause of going light. These may include faecal examinations, blood tests, and X-rays to detect any infections or abnormalities.

Providing a nutrient-rich diet is important for finches experiencing going light. Moving the bird to a hospital cage and offering high-quality seed mixes, fresh greens, multivitamins, electrolytes, and protein supplements can aid in restoring their energy levels. In severe cases, hand-feeding or syringe feeding may be required.

Treatment Guide	
GOING LIGHT	
Treat by elimination of likely cause	
Examples:	1: Coccidiosis
	2: Bacterial infection
	3: Fungal infection

After a diagnosis made, treatment may include antibiotics, antiparasitic, or antifungal medications to treat the underlying cause. Another common approach is to first treat the most probable cause and observe if the bird shows improvement.

Maintaining a clean environment is also important for the recovery of Gouldian finches exhibiting symptoms of going light.

Mutation Health Issues

Gouldian finch mutations are sometimes considered to have health issues. In some cases it is a genetic weakness that is associated with the trait itself - see "Health Problems for Blue Gouldians" on page 91. On other occasions there is a weakness or genetic collapse due to the inbreeding that was used to obtain sufficient numbers of the new mutation. One example of this is Twirling; see "Abnormal Head Movements" on page 172.

New mutations can result in birds that are weaker than the natural "wild-type" varieties. When these birds breed, there is a higher likelihood of producing weaker offspring. As a result, these birds may require more protection from cold draughts and the supply of dietary supplements. See "Special Diets for Special Birds" on page 69.

When pairing mutations these potential health issues must be kept in mind. For instance, the blue mutation is frequently paired with normal split blue Gouldians to produce stronger blue birds.

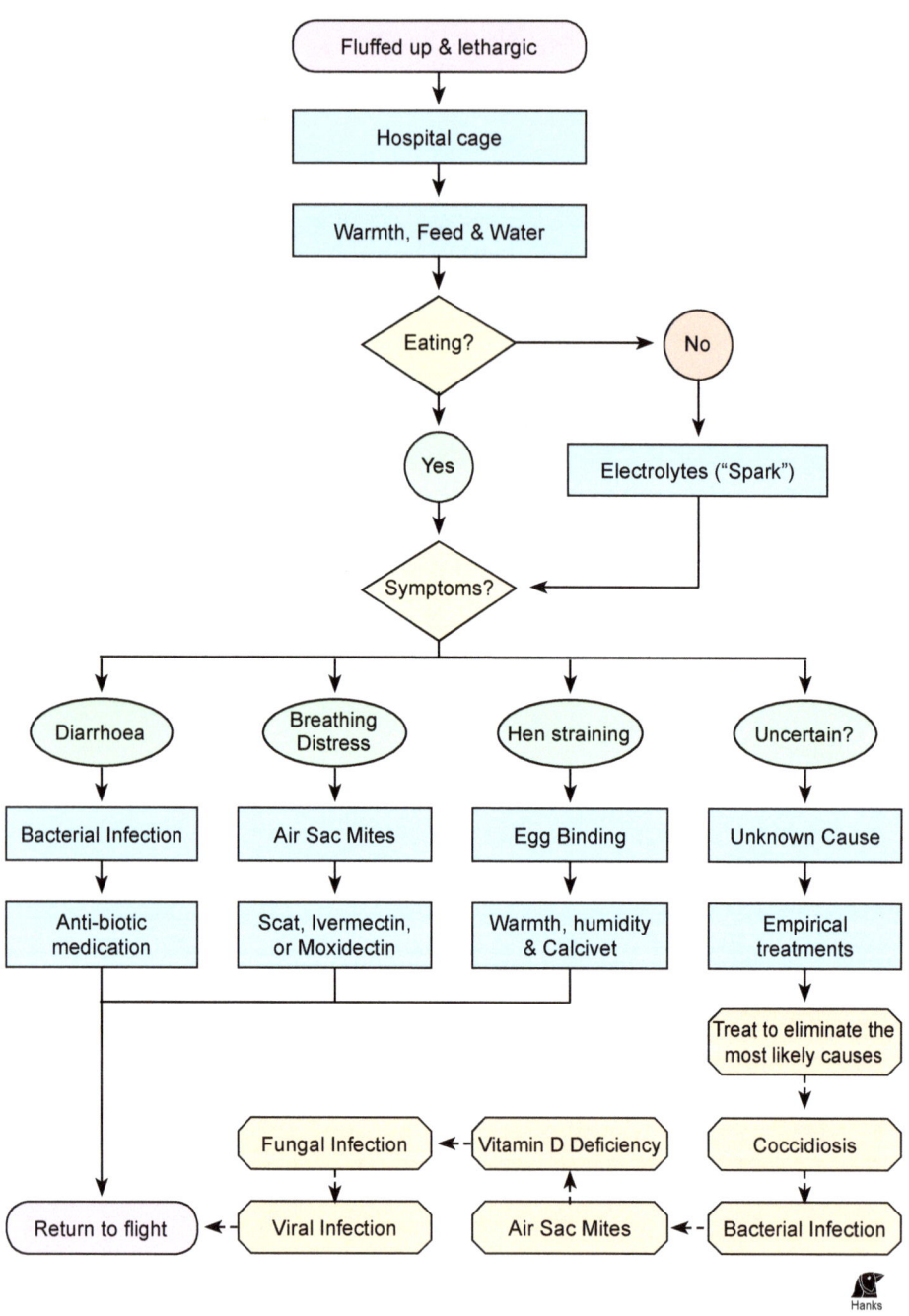

Example of a flow chart for treatment decisions based upon observed symptoms & diagnosis

Health & Wellness

Medications

Gouldian finches, just like other animals, can experience various illnesses and conditions that may require medication. It is important to determine the correct medications and proper administration methods to ensure their safety and effectiveness.

Methods of Administration

Administering medication to aviary birds requires careful attention due to their small size and sensitivity. It is important to follow the instructions provided on the specific product being used.

Oral medications can be applied to the beak of specific birds, but they are more commonly added to the drinking water – either for an individual in a hospital cage or a colony in an enclosure.

Other alternatives include a "spot-on" treatment applied directly to the skin, usually at the back of the head or under the wing; or sprays onto feathers, commonly used for external parasites; or sprinkled onto food.

Dosage Rates

Birds require precise medication dosages. Both overdosing and underdosing can result in significant adverse effects. It is not possible for any publication to provide exact dosage information due to the diverse concentrations of active ingredients and the various product names found in different international markets.

Converting labelled dosage rates into commonly used measurements can be helpful. For instance, certain products specify dosages in terms of drops, ounces, or millilitres, which need to be added to water quantities measured in pints, quarts, millilitres, or litres. A form for preparing such a list is contained in the "Forms Library" on page 203. Examples are also shown in "Popular Medications & Supplements" on page 229.

Remote Dosing

It can be helpful to administer medication without entering the enclosure. This process, known as "Remote Dosing," may be necessary to avoid disturbing breeding birds or if it is not feasible to empty untreated water from a container such as an ornamental bird bath.

When medication is mixed as directed and then introduced into untreated water, the

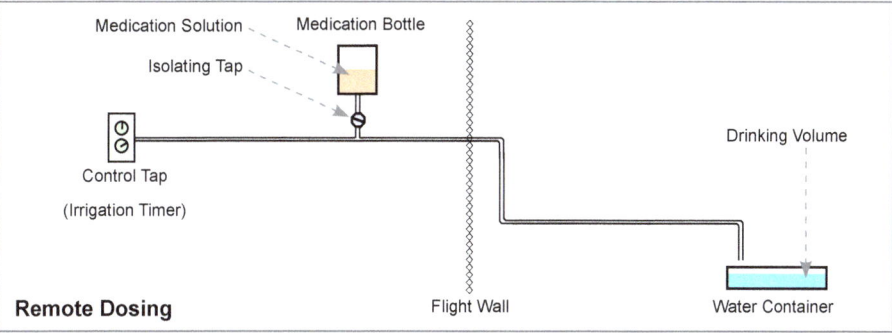

Design example for Remote Dosing in a flight

Remote Dosing Examples						
	Example 1			Example 2		
Drinking Volume	1000ml	(2 x 500ml)		360ml		
Medication Dosage	3ml/1000ml	(0.3%)		3ml/1000ml	(0.3%)	
Medication Required	3ml	(x1.5 = 4.5ml in Med Solution)		1.08ml	(x1.5 = 1.6ml in Med Solution)	
	Water	Medication	Med Conc'n	Water	Medication	Med Conc'n
Fill Fresh Water	1000ml	0ml	0.0%	360ml	0ml	0.0%
Medication Solution	+ 500ml	4.5ml	0.9%	+ 180ml	1.6ml	0.9%
Combine Both	= 1500ml	4.5ml	0.3%	= 540ml	1.6ml	0.3%
Excess Overflows	1000ml	3ml	0.3%	360ml	1.08ml	0.3%

concentration in the combined solution will become diluted and lose its effectiveness. Therefore, it is necessary to perform some basic calculations for proper Remote Dosing:

1. Assuming the water container in the flight is filled by a *Control Tap* outside the flight, add a *Medication Bottle* to that line with an *Isolating Tap* (see diagram).
2. Measure the *Drinking Volume* of the water container in the flight. Calculate the required *Medication Required* for this amount of water.
3. Run the *Control Tap* to completely fill the water container with fresh untreated water.
4. Prepare a jug of concentrated *Medication Solution* that is 0.5x the *Drinking Volume*, but with an increase quantity of medication that is 1.5x the *Medication Required*.
5. Close the *Control Tap* to prevent it from running automatically.
6. Open the *Isolating Tap*.
7. Slowly add the *Medication Solution* to the *Medication Bottle*, using 5 separate pours.
8. The *Medication Solution* will flow into the *Drinking Volume*; mixing the concentration and dispersing in the existing water; while the displaced water will overflow.
9. Note: This dispersion is why the *Medication Solution* needs to be added slowly.
10. Note: When topping up the water on subsequent days, the *Medication Solution* is prepared at the normal concentration.
11. When the treatment period is complete, close the *Isolating Tap*. Then open the *Control Tap* and run it to dilute and flush the water container back to clean fresh untreated water.

Off-Market

Off-market medications are products that are not explicitly licensed or formulated for use in specific animals. While these medications can occasionally fill gaps left by commercially available options, their usage requires additional calculations and responsibilities. For instance, a medication made for poultry would obviously need dosage adjustments for a finch. Another example is Ivermectin for air sac mites, which was originally available for farm animals but is now sold in aviary-specific products. In its early years of use, it was considered "off-market."

When standard treatment options are not available through standard options, off-market medications can provide alternative treatments. One concern with off-market medications is the lack of regulatory oversight, which means there is no assurance of the product's safety or efficacy. The ingredients might cause adverse effects or fail to treat the intended condition. Additionally, without proper veterinary guidance, dosage errors (such as overdosing or underdosing) can lead to significant health issues.

International Products

Many active ingredients are marketed under various product names, particularly across different countries and regions. Simply as an example, the table below offers a comparison of some of these names. However, readers should always conduct their own research to determine the most suitable medications available in their own market.

Examples..	Coccidiosis	Air Sac Mites	Anti-Biotic	Anti-Fungal
Australia	• Coccivet • Cocciprol	• Moxivet Plus • S76	• Avicycline C • Triple C	• Nilstat • Fluconazole
South Africa	• Virbacox • Coccivet	• Masters Moxidectin • TPH Moxidectin	• Phenix • Avivet	• Itraconazole • Revive
UK	• Coxoid • Coccicare	• Meditech Moxidectin • Scatt	• Baytril • Tetracycline	• Fungitraxx • NPR Avian Antifungal
USA	• Cocci Powder • Endocox	• Cocci Powder • Endocox	• Cocci Powder • Endocox	• Fungistat • Medistatin

Safety Precautions

Proper administration of medications is crucial for the health and safety of birds. If a carer is uncertain, they should consult a Veterinarian, preferably an Avian Veterinarian. The outcome of any treatment depends almost entirely on an accurate diagnosis.

Quick Reference Table - Summary of Diseases

This table on page 86 provides a summary of common health conditions in Gouldian finches. Brand names are included, but dosages are not listed due to varying concentrations available in different markets globally. For example, a 2% concentration in one country will have a different recommended dose than a 3% concentration in another. Therefore, it is important to always follow the instructions provided on the product being used.

A record of concentrations for the most common medications can be very useful. See "Popular Medications & Supplements" on page 229 and "Dosage Rates" on page 207 (in the Forms Library).

There is also an example of decisions in the "Flow Chart for Disease Treatment" on page 82.

Quick Reference Table - Diseases of Gouldian Finches

Condition	Symptoms	Cause	Treatment	Example
Respiratory Infection	• Lethargy • Respiratory distress • Nasal discharge	• Bacterial • Fungal • Viral	• (See Medications below) • Electrolytes • Hospital cage	• (See infections below)
Bacterial Infection	• Lethargy • Diarrhoea (Yellow) • Weight loss • Sinus swelling • Smelly droppings	• Bacterial (eg E. Coli, Salmonella, Psittacosis)	• Antibiotic • Electrolytes • Hospital cage	• Tetracyclin • Neomycin • Baytri • Inca Sulpha 3 • Vetafarm Triple C • Spark / Quik Gel
Fungal Infection	• Lethargy • Respiratory distress • Foul smell • Skin infections	• Fungal • Wet enclosure	• Anti-Fungal • Dryer enclosure • Hospital cage	• Nystatin • Itraconazole
Viral Infection	• Lethargy • Diarrhoea (Green) • Respiratory distress	• Viral • Contagious	• Electrolytes • Hospital cage	• Spark or Quik Gel
Mite Infestation	• Respiratory distress • Coughing • Wheezing • Breathing clicks	• Air Sac Mites • Mites	• Mite medication: - in all water supplies *or* - spot on	• Ivermectin • Moxidectin • Scat • S76 • Moxivet
Feather Loss	• Feathers missing • Bald areas	• Stress • Vitamin deficiency • Parasites	• Remove stress • Improve diet • Treat parasites	• Vitamin D3 • Spray parasites • Wait for annual moult

Table Continued →

Health & Wellness

Quick Reference Table - Diseases of Gouldian Finches

Condition	Symptoms	Cause	Treatment	Example
Nutritional Deficiencies	• Feathers missing • Feathers poor condition • Breeding results poor • Weak bones	• Deficiencies in vitamins or minerals	• Better diet • Vitamin supplements • Avoid rancid cod liver oil (Vit E defic.)	• Solaminovit • Ornithon
Vitamin D Deficiency	• Lethargy • Breeding results poor • Growth of young birds poor • Feathers brittle	• Lack of natural sunlight including UV	• Sunlight • Artificial source of UV • Vitamin D3 supplement	• Growth lights • Vetafarm D • Calcivet
Coccidiosis	• Lethargy • Diarrhoea severe (blood) • Weight loss • Eating constant	• Parasite • Damp environment • Wet floor • Common after cold damp conditions	• Environmental cleanliness • Anti-coccidial medication	• Coccivet • Baycox 5% • Carlox • Probiotics
Giardia	• Lethargy • Diarrhoea • Weight loss • Feather plucking	• Parasite • Spreads in water	• Sterilisation of baths & drinkers • Antiparasitic medication	• Ronidazole • Ronivet • Harkanker
Melanism	• Lethargy • Pigments black or brown increased • Feathers brittle • Immune function lower	• Lack of natural sunlight including UV	• Sunlight • Artificial source of UV • Vitamin D3 supplement)	• Growth lights • Vetafarm D • Calcivet, DufoPlus, or Zade
Egg Binding	• Lethargy • Respiratory distress • Straining • Swollen abdomen • Sitting horizontal	• Hen too young or old • Obesity • Calcium defic. • Lighting cycle • Possible infection	• Warmth & humidity • Oil on vent • Hospital cage • Calcium to beak • Calcium in diet	• Olive oil • Calcivet • Zolcal-D

Another useful resource in the Forms Library for arranging thoughts is the "Decision Tree for Health Diagnosis" on page 206.

Use of a Hospital Cage

Design

These isolation cages should be large enough to allow the bird to move comfortably but not so large that it risks further injury by flying or jumping. Typically, a hospital cage of around 36 x 25 x 25 cm (12 x 10 x 10 in) is suitable for Gouldian finches.

Proper ventilation is provided with mesh panels while also ensuring there are no draughts. An adjustable heating element maintains a temperature of around 25°C (77°F) and any lighting is soft to reduce stress and provide a natural day-night cycle.

The bird will likely spend a lot of time on the floor, but a perch helps to encourage return to normal behaviour. Lining the floor with absorbent paper towels is useful for monitoring droppings.

Food & Water

Food and water dispensers should be accessible and positioned to minimize spillage and contamination. Depending on the bird's condition, special feeders or drip water systems, such as Edstrom water valves, may be required.

Advantages

The main function of a hospital cage is to offer a regulated environment that facilitates

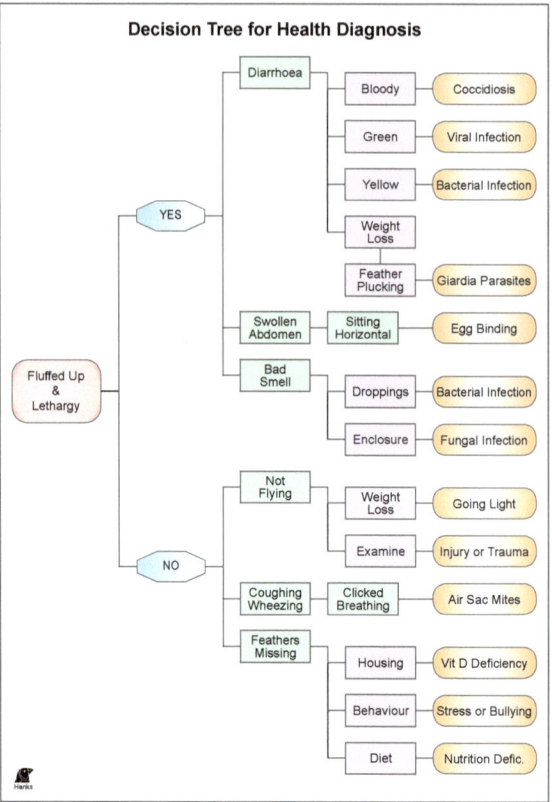

A "Decision Tree" is a useful aid in making the diagnosis

Hospital cage

Health & Wellness

the bird's recovery. The primary advantages include:

- Controlled Environment: Temperature, humidity and lighting are maintained at optimal levels in a clean space.
- Isolation & Stress Reduction: Keeping a sick bird isolated helps prevent the spread of disease and reduces stress.
- Monitoring: A hospital cage enables close monitoring of the bird's condition, behaviour and response to treatment.

The cage should be placed in a quiet low-traffic area that also avoids draughts, direct sunlight or over-heating. After use it is thoroughly cleaned and disinfected to be ready for next time.

Treatment Notes

Every Gouldian in a hospital cage is not only an opportunity to restore the health of the bird, but also to learn for future cases with similar symptoms. To do this it is essential to keep records of the medications that worked, together with the dosage rates and duration of treatment. A form for doing this is provided in the "Forms Library" on page 203.

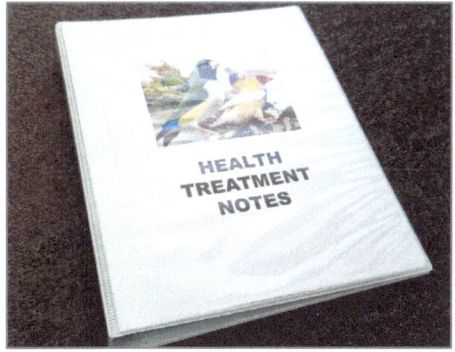

Every health problem is a learning opportunity

Preventive Care

Regular health checks, providing a clean environment and a balanced diet are key to preventing illnesses in Gouldian finches.

Juveniles removed to a holding flight

Proper health care ensures that they are less susceptible to diseases and can recover more quickly if they do fall ill. Ensuring their health through preventive care and a clean, stimulating environment not only improves their quality of life but also supports their natural life cycle and breeding success.

Routine Annual Medications

Obviously some medications, like antibiotics, are used in response to the current circumstances. However, some other medications and supplements should be used as part of a routine health program for Gouldian finches:

Multivitamin	Supplement	Daily	
Calcium	Supplement	Weekly	Or diet. Especially during breeding.
Iodine	Liquid Iodine	Weekly	Or diet supplement.
Electrolytes	Spark	Three Monthly	
Insect protection	Coopex	Three Monthly	Spray enclosure.
Anti-Coccidiosis	Baycox	Six Monthly	Or following severe cold & damp.
Anti-Mites	Moxidectin	Six Monthly	
De-Worming	Moxidectin	Six Monthly	
Nest preparation	Coopex	Annual	Dusting in nest boxes.

For nest preparation the dusting with Coopex powder is placed into the nest box before the initial packing with nesting material. Iodine can be supplied in the form of medicated shell grit and calcium in the form of baked eggshells or cuttlebone.

Routine Trimming of Claws

An often overlooked aspect of caring for birds is the trimming of claws, but this procedure is essential for the bird's health and safety.

Bird claws consist of a hard keratin outer shell and a softer sensitive core called the quick, which contains blood vessels and nerves. In their natural habitat, birds wear down their claws through activities like perching on rough surfaces. In aviaries, Gouldians may not have the same opportunities for natural claw wear, which can result in overgrown claws.

Once overgrown, the claws can grow into the bird's footpads, causing infections and injuries. Additionally, long claws can get caught in cage bars, potentially leading to broken or torn claws, which may then be prone to infection.

To trim a bird's claws, gather the necessary tools and be in a calm

Tools for claw trimming

environment. Required tools include specialized bird claw trimmers, or human nail clippers, and styptic powder to stop bleeding if accidental cuts occur.

> **TIP** If uncertain, it is better to trim claws by a lesser amount and repeat the procedure more often.

Inspect each claw to identify the quick, which is visible as a pinkish area inside the claw. Trim only the tip of the claw, avoiding the quick to prevent bleeding and discomfort.

Health Problems for Blue Gouldians

When considering the Blue Gouldian mutations it is common to be told that they are hard to keep alive; that they need to be given a dietary supplement; or that they need only the same nutrition as wild-type birds. To explain the health aspects of Blue Gouldians it is necessary to first understand the mechanism that makes them appear blue: This is the reduced processing of carotenoids.

Carotenoids are natural pigments that birds obtain from their diet, mainly through plants, insects and seeds. These compounds produce red, orange and yellow colours; so when carotenoids are reduced, colours such as green will appear blue.

In addition to influencing plumage colour, carotenoids play an important role in antioxidant protection, immune function, vision and reproductive health. They help neutralize free radicals and serve as precursors to producing vitamin A – a nutrient that is essential for maintaining epithelial tissues and enhancing immune system performance.

Therefore the lack of carotenoids, which results in the blue colouring, can also limit the production of vitamin A. In some species, a blue mutation still allows for carotenoid absorption and the health benefits, even though these pigments are not expressed in their plumage. For instance, this occurs with the dominant white (blue) mutation in canaries.

However, in the case of Blue Gouldians they simply do not absorb carotenoids normally, so there are health consequences. This is why extra dietary carotenoids (lutein, zeaxanthin and cryptoxanthin) like broccoli, yellow corn and pumpkin are recommended; and they do actually need Vitamin A supplements, because carotenoids are far more than mere pigments.

For more information . . .
- Special diets - see "Supplements for the Blue-Backed Mutation" on page 69.
- Feather colour - see "The Science of Gouldian Colours" on page 163.
- Genetics - see "Blue-Backed Mutation" on page 125.

Apple Cider Vinegar

Apple Cider Vinegar (ACV) is a vinegar made from fermented apple juice. It contains acetic acid, vitamins, minerals and enzymes that have antimicrobial, antifungal, and antioxidant properties. It can be used as a natural remedy for Gouldian finches.

Benefits of ACV

Apple Cider Vinegar can be a simple yet effective addition to the health care of finches:
- Treatment of bacterial, fungal and viral infections.

- Improved Digestion and Gut Health - a lower pH level and promoting the growth of beneficial bacteria while inhibiting harmful pathogens. This can lead to improved nutrient absorption and reduced risk of digestive issues.
- Boosted Immune System - making the birds more resilient to infections and diseases by acting as a natural antibiotic.
- Better Respiratory Health - ACV reduces the risk of respiratory infections and the anti-inflammatory effects alleviate symptoms of respiratory distress and lower mucous.

Apple Cider Vinegar diluted for use as a safe disinfectant

- Controlling Parasites - the acidic nature of ACV creates a less attractive environment for external parasites, such as mites and lice.
- Increased Egg Production - improved shell strength and better regulated calcium levels.

Selecting the Right ACV

Using raw, unfiltered "organic" Apple Cider Vinegar that contains "the mother" is recommended because it includes beneficial bacteria and enzymes. Processed or distilled vinegars typically do not contain these beneficial components.

Guidelines for Use of ACV

Apple Cider Vinegar can be used as a general disinfectant that is safe for birds. Diluted at 15% in water it can be sprayed onto cage and cabinet surfaces during routine cleaning.

When adding Apple Cider Vinegar to drinking water, it is important to dilute it properly to prevent any potential harm. A general guideline for safety and effectiveness is 5 to 10ml per litre of water (approximately 1 teaspoon of ACV per quart of water - which converts to 5ml in 946ml). The suggested frequency of use for finches is once per week, and this should be done consistently.

Treatment Guide	
GUT HEALTH	
Supplement in drinking water	
Examples:	Organic ACV
Duration:	1x / week

The precautions include diluting apple cider vinegar (ACV) before use; introducing it gradually to allow birds to adjust; monitoring for any adverse reactions; and consulting with an avian veterinarian if necessary.

When to Consult a Veterinarian

Understanding the signs that indicate when to seek professional help is crucial. This includes changes in behaviour, appetite, or physical appearance. Remember that useful treatment depends upon accurate diagnosis. If you are uncertain, it is good advice to consult an expert.

Fractures and trauma will generally need the assistance of a veterinarian in order to make an accurate diagnosis and set the fracture if appropriate.

Other times to consult a veterinarian is when there are signs of infection and a treatment needs to be prescribed, or faecal pathology examinations can help early identification of infections and limit outbreaks.

Standard Care Plan for Gouldians

	☐ MAINTENANCE	☑ BREEDING / MOULTING				Date:	10/10/2025	
	Product Name	Dosage	Days of 7	Vitamin A	Vitamin D	Vitamin E	Calcium	Iodine
Water	Ornithon	4g	1	570	57	1.3	0	0
Dry Seed	Soluvite D	10g	7	10,000	2,500	25	0	2,000
Wet Food	The Good Oil	15ml	2	214	171	10	0	0
	D Nutrical	5g	2	714	71	1	442	0
			Recomnd:	M5/B11000	M1/B3000	M10/B50	M2/B5000	M180/B180
			Totals:	11,498	2,799	37.3	442	2,000

	Product Name	Days of 7	Product Name	Days of 7	Product Name	Days of 7
Extras	☑ Grit: Fit Grit	7	☑ Cuttlebone	7	☑ Baked Eggshells	7
	☑ Greens & Grains	1	☑ Seeding Grasses	3	☐ Direct Sun >30min	
	☑ Wet Fd: Soak Sd	2	☐ Live Fd: _____		☐ _____	
	☐ Spinach ☑ Kale		☑ Chickwd ☐ Endive ☑ Sprouts ☐ Cucmbr		☐ _____	2

	Purpose	Product Name	Days of 7	Mths of 12	Dosage	Notes
Routine Meds	☑ Coccidiosis	Baycox		3	3ml	After cold / wet
	☑ Air Sac Mites	Moxidectin Plus		6	5ml	Repeat 14 days
	☑ Worms	" "		"	"	" "
	☑ Nest Preparation	D Earth		12	3g	Dust under grass
	☑ Mites & Insects	Coopex		4	5g	Spray
	☑ Water Disinfectant	KD Powder	1		1g	
	☐ _____					

(All doses are per L or Kg)

A completed Care Plan as an example only

Quick Reference Table - Measurements

1 litre (L) = 1,000 ml	= 1,000,000 mcL	
1 millilitre (ml) = 1,000 mcL (or µl)		
1 microlitre (mcL)		
1 Drop (gtt) = 0.05 ml	= 50 mcL	
1 US pint (pt) = 473 ml	= 473,176 mcL	
1 US cup (c) = 240 ml	= 240,000 mcL	
1 US fluid ounce (fl oz) = 29.6 ml	= 29,573 mcL	
1 US tablespoon (Tbsp) = 14.8 ml	= 14,787 mcL	
1 US teaspoon (tsp) = 4.9 ml	= 4,928 mcL	
1 Imperial pint (pt) = 568 ml	= 568,261 mcL	
1 Imperial cup (c) = 284 ml	= 284,131 mcL	
1 Imperial fluid ounce (fl oz) = 28.4 ml	= 28,413 mcL	
1 Imperial tablespoon (Tbsp) = 17.8 ml	= 17,758 mcL	
1 Imperial teaspoon (tsp) = 5.9 ml	= 5,919 mcL	
1 kilogram (Kg) = 1,000 g	= 1,000,000 mg	= 1,000,000,000 mcg
1 gram (g) = 1,000 mg	= 1,000,000 mcg	
1 milligram (mg) = 1,000 mcg		
1 microgram (mcg)		
1 pound (lb) = 454 g	= 453,492 mg	
1 ounce (oz) = 28.3 g	= 28,349 mg	
1 metre (M) = 100 cm	= 1,000 mm	= 1,000,000 mcm
1 centimetre (cm) = 10 mm	= 10,000 mcm	
1 millimetre (mm) = 1,000 mcm		
1 micrometre (mcm)		
1 foot (ft) = 30.5cm	= 304.8 mm	
1 inch (in) = 25.4 mm	= 25,400 mcm	

Sometimes the quantities in instructions need to be converted

6

BREEDING GOULDIAN FINCHES

Breeding Gouldian finches can be a captivating and rewarding experience for both novice and experienced aviculturists. These colourful birds are bred for their aesthetic appeal as well as for conservation.

There is also a personal satisfaction in breeding the best possible examples of these beautiful birds and breeders are often aiming to enhance and maintain the genetic diversity within their birds.

Importance of Genetics

Breeding Gouldian finches allows for the study of avian genetics. A basic understanding of genetics is necessary to achieve better pairings. In normal wild-type birds, this knowledge helps in obtaining good health and show quality. While in mutation variations, it is genetics that provides the opportunities for producing different colour combinations.

Gouldian genetics are discussed in the chapter titled "Genetic Considerations". See page 119.

Understanding the Breeding Cycle

To successfully breed Gouldian finches, aviculturists must first understand their natural breeding cycle. As with most species, the breeding season typically begins when the conditions in their environment are optimal. In nature this is when daylight hours are longer and there is an ample supply of grass-seeds. These conditions trigger hormonal changes in the birds, preparing them for mating.

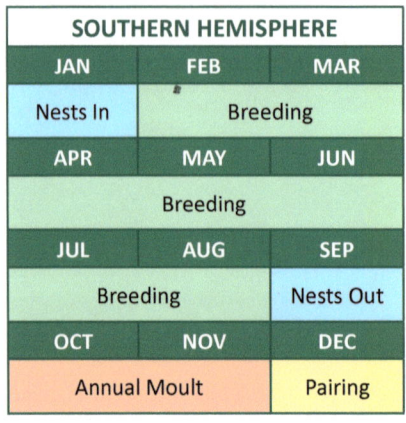 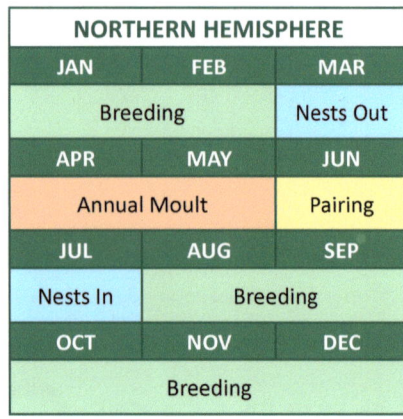

Typical breeding timetables for Gouldian finches

Annual Timetable

Gouldian finches originate from north-west Australia in the Southern Hemisphere, where the longest days of summer occur in December. They begin pairing in November and December; followed by nest building in January (the "Wet Season"); breeding from February to August; and then the annual moult in September and October.

This timetable is reversed when Gouldian finches are living in the Northern Hemisphere.

Colonies or Pairs

Gouldian finches can be bred either in colonies or pairs. Both methods come with specific advantages to consider.

Breeding Gouldians in Colonies

Breeding Gouldian finches in colonies involves housing multiple pairs together in a single flight. This method allows for social interaction in an environment similar to their natural habitat, where they typically live in flocks.

The competitive and stimulating environment of a colony can enhance breeding success, while the presence of multiple nests provides a sense of security that encourages breeding. Additionally, a colony flight can be more efficient by reducing the need for numerous smaller cages.

Breeding Gouldians in Pairs

Breeding Gouldians in pairs involves housing a single pair in a dedicated breeding cabinet. This method allows for better control of the breeding environment and closer monitoring of the health, diet, and behaviour of each pair.

While finches are typically social, some individuals may show aggression in a colony setting. Breeding in pairs minimizes the risk of aggressive encounters, particularly for Red Headed Gouldians, which tend to be more aggressive in colonies. However, the biggest advantage of breeding in pairs is that it allows for selective breeding, especially when aiming for specific genetic traits or mutations.

Although colony and pair breeding methods have distinct advantages, both should be considered with the expected progeny in mind. For instance, it is logical to breed standard Black Headed Gouldians within a colony setting, whereas the various Gouldian mutations would be more effectively understood through pair breeding.

Creating the Optimal Breeding Environment
Preparation for Breeding
Proper preparation involves selecting healthy breeding pairs, providing a suitable nesting environment, and ensuring nutritional readiness. Selecting pairs is discussed later and nutrition was discussed earlier.

1. Cocks and hens were already in separate flights for the annual moult after the previous breeding season.
2. As juveniles colour up and they are moved into the appropriate single-gender flights.
3. Any excessive stock is moved on and new blood lines are added from other breeders.
4. All breeding cabinets, hoppers, trays and utensils are thoroughly cleaned and washed. See "Annual Cleaning" on page 74.
5. All nests are cleaned and then disinfected with an aviary disinfectant (eg Vetafarm "Avi-care" or a 15% dilution of Apple Cider Vinegar.
6. Austerity feed is provided to help to simulate the natural cycles and encourage breeding patterns when the diet is later enriched. See "Austerity Seed" on page 52.
7. Intended pairings are planned based upon progeny predictions and avoiding the confusion of normal and split birds occurring in the same clutches. See "Breeding Plans Based on Predicted Progeny" on page 103.
8. To allow for bonding, birds are paired in breeding cabinets 2 to 4 weeks before it is planned to install nest boxes.
9. To protect the birds from insects and mites, the nest boxes are sprayed with Coopex, dried and then dusted with Diatomaceous Earth.
10. Nest boxes are installed with a basic nest formed in each one and nesting material is provided in each cabinet or flight.
11. Baked eggshells are already part of the standard nutritional offering.
12. As part of the "Wet Season" stimulation of breeding this is also the time when fresh

A breeding cabinet with nest box attached

seeding grasses are offered every second day, together with a separate supply of Greens & Grains seed.

Suitable Nest Boxes

When choosing a design for a Gouldian nest box, it is important to create an environment that mimics their natural nesting sites. These nest boxes are usually made from materials such as plywood, chipboard, plastic, or compressed foam, all of which meet the birds' requirements.

Availability of seeding grasses is a natural trigger to breed

The main part of the nest box should be approximately 16 cm long, 15 cm high, and 13 cm wide internally, providing adequate space for the birds to nest comfortably. One of the distinctive features of a Gouldian nest box is a second entry through a veranda. This additional entryway offers the birds increased privacy and helps reduce stress. Additionally, the nest chamber is slightly lowered, which mimics the dark hollows in trees that Gouldians prefer in the wild. This design feature also helps to prevent chick-tossing out of the nest, a behaviour that can occur under some circumstances discussed later under "Chick Tossing" on page 167.

Preparation of Nest Boxes

First spray each nest box with a bird-safe insecticide like Coopex (pyrethrin). Allow to dry, then sprinkle Diatomaceous Earth into each nest box to further discourage insects

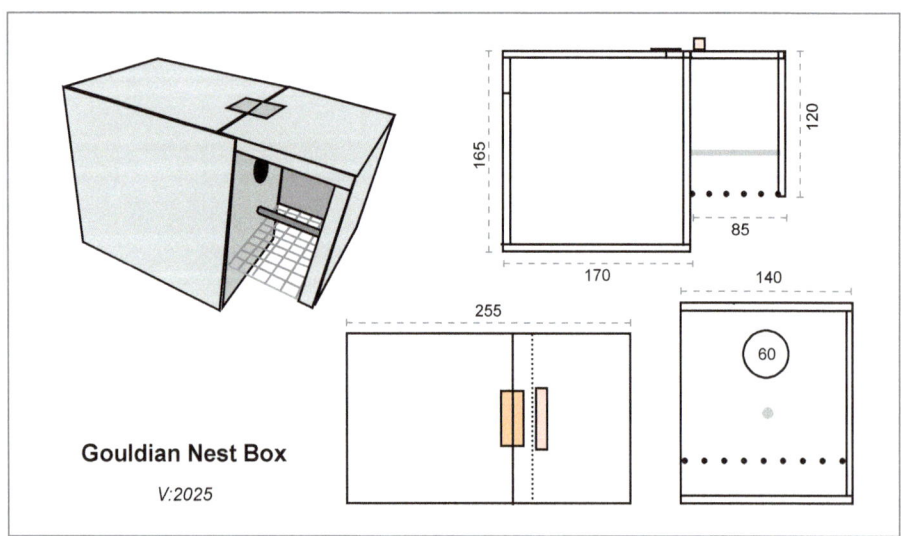

A nest box designed specifically for Gouldian finches

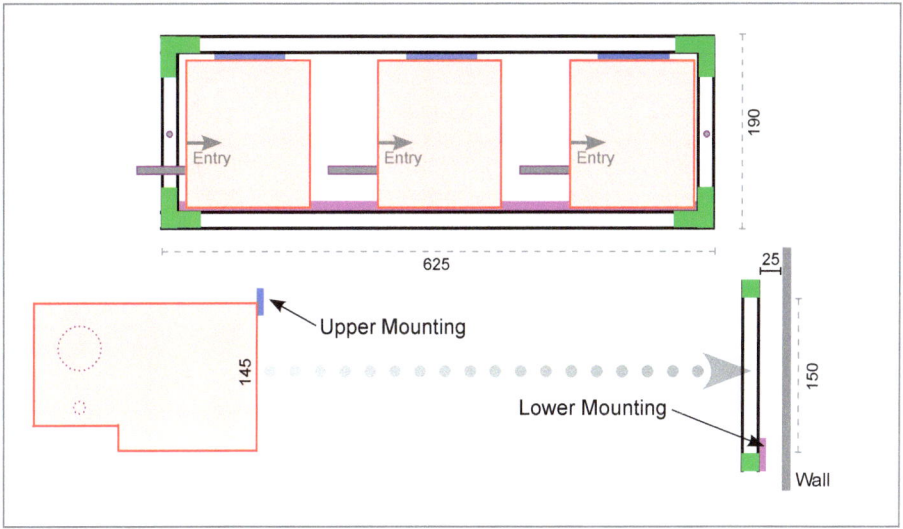
Design for a nest frame for a flight where nests are held in place by their own weight

like mites. (See page 175). Gouldian finches may sometimes not be dedicated to nest building; thus it is advisable for breeders to partially shape the nest when installing them. The nest box should be lined with soft nesting materials such as dried grass or coconut fibre to initiate the nest construction. This approach will further encourage the birds towards successful egg-laying and chick-rearing.

Installation of Nest Boxes

When nest boxes are installed in flights or breeding cabinets it is preferable to use a system that combines holding each nest securely, while also making it easy to remove for inspections and annual cleaning.

The best way to achieve this is with a system where the nest is simply held in place by its' own weight. This is referred to as a Nest Frame for flights, or a Nest Mount for breeding cabinets.

Pairing and Mating: Selecting Compatible Pairs

Breeding Gouldian finches starts with the selection of breeding pairs. Only Gouldian finches who have completed the annual moult are considered.

It is important to choose healthy birds to produce sturdy offspring. Birds with vibrant colours and well-formed bodies are typically selected, as these traits are often inherited by their progeny.

Close Relatives

Genetic diversity is a crucial aspect to consider. By pairing finches with varied genetic backgrounds, it is possible to minimize the risk of hereditary health problems and improve the overall vitality of the offspring. Therefore, breeding close relatives should be avoided. For instance, siblings or a mother and her son.

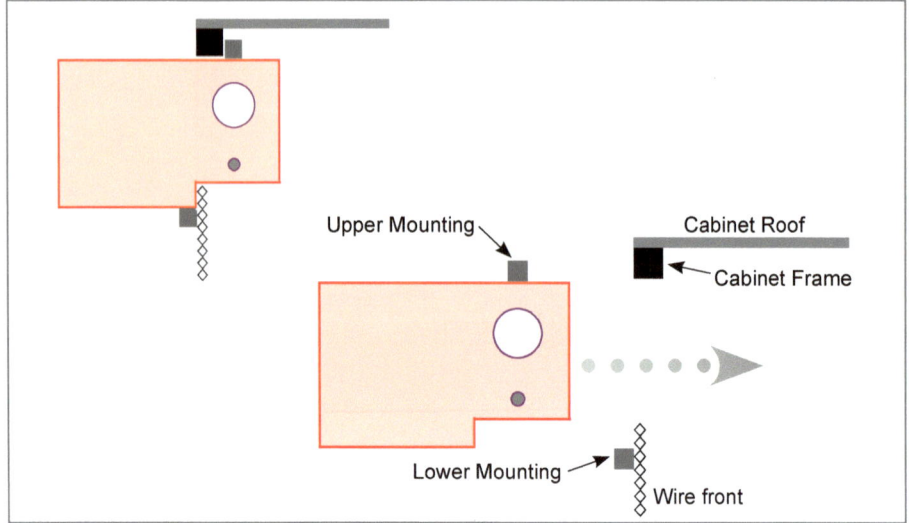
Design for a nest mount in a cabinet where nests are held in place by their own weight

Ages

Gouldian finches should be at least 12 months old before being introduced into breeding situations with nests available. Younger hens are more susceptible to egg binding, and both sexes tend to be more effective parents when they are older. Hens over 5 years old may have reduced fertility.

Head Colours

In the wild, Gouldian finches tend to breed with other birds of the same head colour. Red-headed varieties rarely choose to breed with black-headed ones. It is generally not advised to select pairs with different head colours for breeding as this can increase the chances of genetic anomalies and dilute the colour vibrancy or purity over generations. This practice may result in offspring with less distinct and more muddled colour patterns. Preserving clear and vibrant colouration helps maintain the Gouldian finch's visual appeal and genetic integrity.

There is also evidence to show that hens will produce more cocks when paired to a Gouldian finch of a different head colour. For more information see "Mating Preferences Related to Head Colour" on page 115.

When Gouldian finches breed in colonies, they are typically separated by head colour for the same reasons. One colony will have black-headed finches, another will have red-headed finches, and a third will have yellow-headed finches.

Pair Bonding

Pair bonding in Gouldian finches is the process of forming a close and long-term relationship between a cock and a hen, which is important for successful breeding. This bond includes cooperation in activities like nest building, incubating eggs and raising chicks.

Formation of Pair Bonds

The formation of pair bonds in Gouldian finches starts with courtship behaviours. Cocks engage in displays to attract females, such as singing and performing a distinctive dance. Hens respond to these displays and select a mate based on the quality of the display as well as the health and vitality of the male.

Once a pair bond is established, the cock and hen show increased attentiveness towards each other, engaging in activities that strengthen their connection and prepare them for breeding. Therefore, it is important to allow adequate time for the bond to develop before introducing nest boxes. A minimum period of 3 or 4 weeks is a good guideline.

Indicators of a pair bond development include activities such as mutual preening, where they clean each other's feathers, and other synchronized activities that serve as a communication and coordination between the pair.

> **TIP** As a key to a successful breeding program for Gouldian finches, allow at least 3-4 weeks for pair bonding before nest boxes are introduced..

Impact on Breeding Success

Strong pair bonds enhance reproductive success through several mechanisms:

- Greater Cooperation
- More Parental Investment
- Improved Offspring Fitness & Survival

Challenges to Pair Bonding

Despite the benefits, Gouldian finches encounter various challenges that can affect pair bonds or hinder their formation. Factors such as environmental stress, competition for mates and habitat disturbances can all impact the stability of these bonds.

→ Non-Breeding Season - When Gouldians are separated by gender during the non-breeding season, it is advisable to also prevent cocks and hens from seeing each other. This helps to avoid the formation of pair bonds before the birds are selected based on the intended breeding plan.

→ Breeding Cabinets - Similarly, for Gouldians that are already paired in breeding cabinets, it is better if they cannot see other potential mates in cabinets opposite.

→ Colony Flights - In colony breeding flights, an imbalance in the sex ratio can lead to increased competition for mates and this may result in unstable pair bonds. So ensuring balanced sex ratios is important for reducing competition and maintaining strong pair bonds.

Predictable Progeny

For selected breeding in cabinets or flights, it is always useful to consult the genetic records of the birds being paired. This information aids in making informed decisions to ensure that the chosen pairs are genetically compatible. It is advisable to choose pairings where there is complete confidence about the predicted outcomes, or progeny.

Breeding Plan Based on Predicted Progeny

Cage Nbr	1		Cage Nbr	6
Description	YH Cock & BH Hen		Description	Pastel Blue & Blue
Cock	Yellow Head Normal		Cock	Red Head Pastel Blue
Hen	Black Head Normal		Hen	Black Head Blue
Progeny Expected	Cocks: RH split BH/YH Hens: RH split YH		Progeny Expected	Cocks: RH Blue, RH Pastel Hens: RH Blue, RH Silver
Cage Nbr	2		Cage Nbr	7
Description	Dilute Cock & YB Hen		Description	Yellow Back & Normal
Cock	Red Head Dilute (SF YB)		Cock	Yellow Head Yellow Back
Hen	Black Head Yellow Back		Hen	Black Head Normal
Progeny Expected	Cocks: RH YB, RH Dilute Hens: RH YB, RH Normal		Progeny Expected	Cocks: Red Head Dilute Hens: Red Hd Yellow Back
Cage Nbr	3		Cage Nbr	8
Description	BH White Breasted		Description	Seagreen & Normal
Cock	Black Hd White Breasted		Cock	Red Head Seagreen
Hen	Black Hd White Breasted		Hen	Red Head Normal
Progeny Expected	All BH White Breasted		Progeny Expected	Cocks: Red Head Normal Hens: Red Head Seagreen
Cage Nbr	4		Cage Nbr	9
Description	Silver Cock & Blue Hen		Description	Normal & Split Blue
Cock	BH Silver (YB YB & bk bk)		Cock	RH Normal
Hen	BH Blue		Hen	RH Normal Split Blue
Progeny Expected	Cocks: BH Pastel Blue Hens: BH Silver		Progeny Expected	Cocks & Hens: RH Normal & RH Normal Split Blue
Cage Nbr	5		Cage Nbr	10
Description	Yellow Back		Description	Split Aust Yellow x Split
Cock	RH Yellow Back		Cock	BH Normal Split AY
Hen	BH Yellow Back		Hen	BH Normal Split AY
Progeny Expected	Cocks: RH Yellow Back Hens: RH Yellow Back		Progeny Expected	Cocks & Hens: BH Normal, Normal Split AY & Aust Yel.

Example of a Breeding Plan for known outcomes - #9 & #10 have uncertain outcomes, so they would not proceed

The Benefit of Predicting

For example, assuming there are two blue-backed birds and two split blue-backed in a collection with normal wild-type Gouldians. Consider the five possible pairings and their progeny predictions . .

	Blue-Backed	Split Blue-Backed	Normal
Blue-Backed	**1.** • 100% Blue-Backed	**2.** • 50% Blue-Backed • 50% Split Blue-Backed	**3.** • 100% Split Blue-Backed
Split Blue-Backed		**4.** • 25% Blue-Backed • 50% Split Blue-Backed • 25% Normal	**5.** • 50% Split Blue-Backed • 50% Normal
Normal			**6.** • 100% Normal

Example of a progeny prediction from possible pairings

Many of the offspring in these examples will look like Normal phenotypes, but some could be split for Blue-Backed in the genotype? Only those 4 pairings shaded green will have known genotypes. The pairings in examples 4 & 5 should therefore be avoided. Otherwise the breeder will have produced birds of normal appearance where the genotype is unknown, but some split and some are normal.

By carefully selecting breeding pairs, breeders can produce offspring with desirable traits such as vibrant colours, robust health and specific mutations.

Breeding Plans Based on Predicted Progeny

Creating a Breeding Plan is the best way to avoid situations where the breeder is uncertain about whether a normal looking bird is actually carrying a split gene for a specific trait. Essentially, the aim is always to produce birds where the genotype is known.

The actual progeny predictions are summarised in the table titled "Quick Reference Table - Gouldian Finch Progeny" on page 150, or available online by using genetic calculators.

Using this information, an example of a Breeding Plan is shown on the following page. The form for preparing a breeding plans is included in the "Appendices" of this book. See "Forms Library" on page 203.

Example of a Non-recommended Pairing

By predicting the possible progeny, a Gouldian breeder can avoid outcomes where the result would have been producing birds with unknown genotypes. It is always better to avoid this confusion so that accurate records can be kept for every bird.

The calculation of the potential progeny is described in the chapter about "Genetic Considerations" on page 119. The following example is an illustration of the confusion that can be produced . .

Example: Red Headed (split Black Hd) cock x Red Headed (split Yellow Hd SF) hen
12.5%
Red Headed
cocks
12.5%
Red Headed
(split Black Head)
cocks
12.5%
Red Headed
(split Yellow Head SF)
cocks
12.5%
Red Headed
(split Black & Yellow
Head) cocks
12.5%
Red Headed
hens
12.5%
Black Headed
hens
12.5%
Red Headed
(split Yellow Head SF)
hens
12.5%
Black Headed
(split Yellow Head SF)
hens

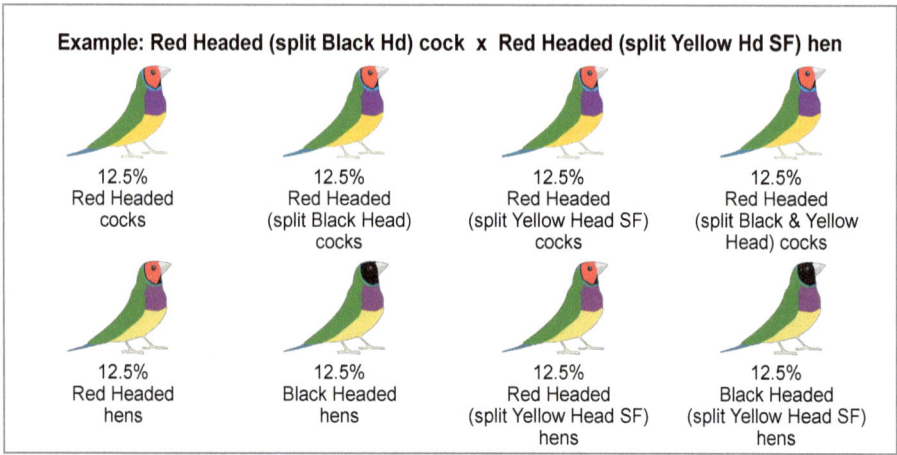

100% of the possible progeny in this pairing would have uncertain genotypes

Examples of Breeding Plans for Mutations

Breeding Plans for Gouldian mutations are based upon understanding the expected progeny predictions using the genetics involved as described in the next chapter of this book. However some suggested pairings for different Gouldian mutations shown here under "Sample Breeding Plans" on page 104 to page 107.

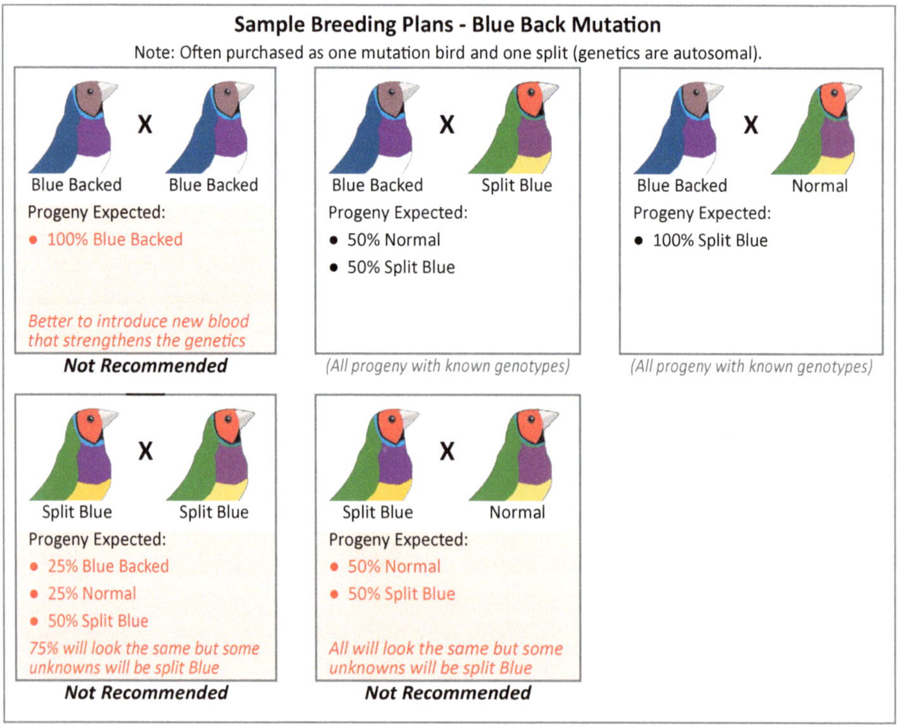

Sample Breeding Plans - Blue Back Mutation
Note: Often purchased as one mutation bird and one split (genetics are autosomal).
X
Blue Backed
Blue Backed
Progeny Expected:
• 100% Blue Backed
Better to introduce new blood that strengthens the genetics
Not Recommended
X
Blue Backed
Split Blue
Progeny Expected:
• 50% Normal
• 50% Split Blue
(All progeny with known genotypes)
X
Blue Backed
Normal
Progeny Expected:
• 100% Split Blue
(All progeny with known genotypes)
X
Split Blue
Split Blue
Progeny Expected:
• 25% Blue Backed
• 25% Normal
• 50% Split Blue
75% will look the same but some unknowns will be split Blue
Not Recommended
X
Split Blue
Normal
Progeny Expected:
• 50% Normal
• 50% Split Blue
All will look the same but some unknowns will be split Blue
Not Recommended

Breeding Gouldian Finches

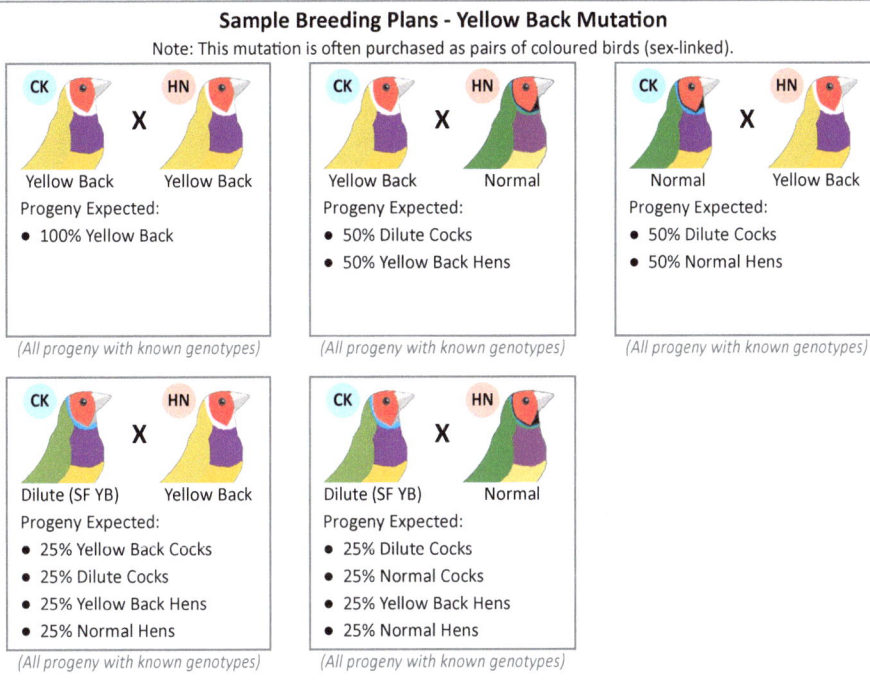

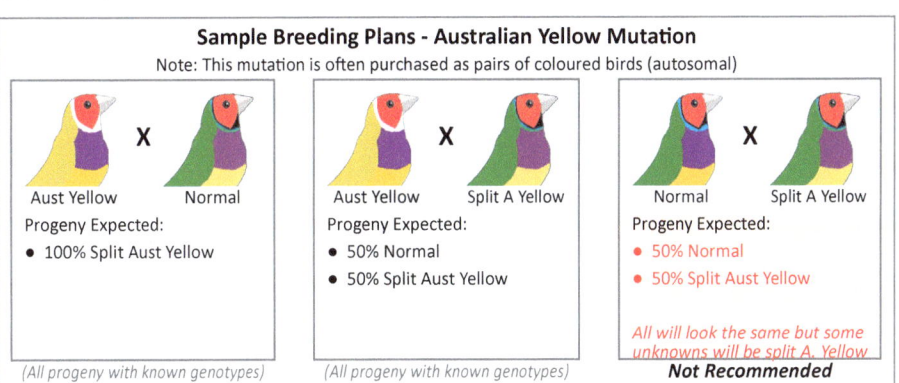

Courtship
Male Performance

Once paired, the Gouldian finches will start to engage in courtship behaviours if they are compatible. Male Gouldian finches perform elaborate courtship rituals to attract a mate. The cocks, distinguished by their bright and varied plumage, perform complex dances and sing melodious songs to appeal to potential female partners. These displays include puffing up their feathers, bobbing their heads and rapid vertical movements on the perch. The vibrant colours of the males and these courtship behaviours are essential in attracting females.

Sample Breeding Plans - Pastel Blue Mutation

Note: Often purchased as one mutation bird and one split (sex-linked Yellow Back and autosomal Blue).

CK Pastel Blue X **HN** Normal
Progeny Expected:
- 25% Normal /Blue Cocks
- 25% Dilute /Blue Cocks
- 25% Normal /Blue Hens
- 25% Yellow Back /Blue Hens

(All progeny with known genotypes)

CK Pastel Blue X **HN** Yellow Back
Progeny Expected:
- 25% Dilute /Blue Cocks
- 25% Yellow Back /Blue Cocks
- 25% Normal /Blue Hens
- 25% Yellow Back /Blue Hens

(All progeny with known genotypes)

CK Pastel Blue X **HN** Blue
Progeny Expected:
- 25% Blue Cocks
- 25% Pastel Blue Cocks
- 25% Blue Hens
- 25% Silver Hens

Better to introduce new blood that strengthens the genetics

Not Recommended

CK Pastel Blue X **HN** Split Blue
Progeny Expected:
- 12.5% Blue Cocks
- 12.5% Normal /Blue Cocks
- 12.5% Pastel Cocks
- 12.5% Dilute /Blue Cocks
- 12.5% Blue Hens
- 12.5% Normal /Blue Hens
- 12.5% Y. Back /Blue Hens
- 12.5% Silver Hens

(All progeny with known genotypes)

CK Pastel Blue X **HN** Y. Back /Blue
Progeny Expected:
- 12.5% Pastel Cocks
- 12.5% Dilute /Blue Cocks
- 12.5% Silver. Cocks
- 12.5% Y. Back /Blue Cocks
- 12.5% Blue Hens
- 12.5% Normal /Blue Hens
- 12.5% Silver Hens
- 12.5% Y. Back /Blue Hens

(All progeny with known genotypes)

CK Pastel Blue X **HN** Silver
Progeny Expected:
- 25% Silver Cocks
- 25% Pastel Blue Cocks
- 25% Blue Hens
- 25% Silver Hens

Better to introduce new blood that strengthens the genetics

Not Recommended

Female Readiness

Readiness of the hen for breeding is shown in the blackened colour of her beak - however this colour change does not happen for the Yellow Back mutation. If she is responsive to the courtship of the cock she faces him, wipes her beak on the perch and tilts her body forward with her tail elevated (possibly quivering).

Know the Difference . .
Gouldian Breeding Season

Hen with Blackened Beak | Hen with no beak change

© 2025 - Hanks, Tony: Gouldian Finches - Care, Breeding & Genetics

The darkened beak colour shows hen readiness to breed

Timetable for Hatching and Independence

Gouldian finches follow a predictable schedule for laying eggs and rearing chicks. Understanding this timetable enables breeders to make informed decisions, such as

whether a nest should be restarted, when sprouted seeds should be offered, when fledglings are anticipated, or when juveniles become independent.

Egg-Laying

This occurs at a rate of one egg per day, with an average clutch size ranging from 4 to 6 eggs. Incubation of the eggs does not commence until after the final egg has been laid, resulting in synchronous incubation and hatching. Newly laid eggs remain viable for up to 7 days prior to the initiation of brooding.

Date Calculations

The typical schedule for Gouldian finches is shown in the "Gouldian Breeding Timetable"

By understanding one event, it becomes possible to predict subsequent events through calculations. For instance, if nestlings are observed to have pin feathers and closed eyes at 6 days old, they can be expected to fledge 18 days later. Similarly, once they leave the nest, they will typically become independent 24 days later.

Incubation

During the incubation period, both parents keep the eggs warm. After hatching, chicks require intensive care. They are altricial, meaning they are born blind and featherless, relying entirely on their parents. The parents provide a diet high in protein and moisture, which is crucial for the chicks' growth and development. This period lasts about 24 days until the chicks are ready to fledge and leave the nest.

Newly hatched chicks

| GOULDIAN BREEDING TIMETABLE ||
EVENT	DAYS
First Egg After Copulation	5 Days
Incubation Period	16 Days
Hatching	0 Days
Pin Feathers	Hatched +6 Days
Eyes Open	Hatched +7 Days
Full feathers	Hatched +15 Days
Fledging	Hatched +24 Days
Independence	Fledging +24 Days Hatched +48 Days

Fledglings

After fledging, young finches depend on their parents for food and protection until they become fully independent. This post-fledging care allows the young birds to develop the skills needed to survive on their own.

Independence

Upon achieving independence, it is advisable to separate juveniles from their parents, particularly if they are housed in a breeding cabinet. This separation enables the breeding pair to return to the nest and produce another clutch without interference from their offspring. While the prediction of 48 days is a general guideline, always verify independence through careful observation before removing juveniles from their parents.

Next Breeding Cycle

Egg laying is a considerable task for hens. When combined with the responsibilities of raising chicks, it can impact bird health. Therefore, the number of breeding cycles per year are generally limited to two. Only in exceptional cases, when one clutch is small, is this extended to three cycles in the same season.

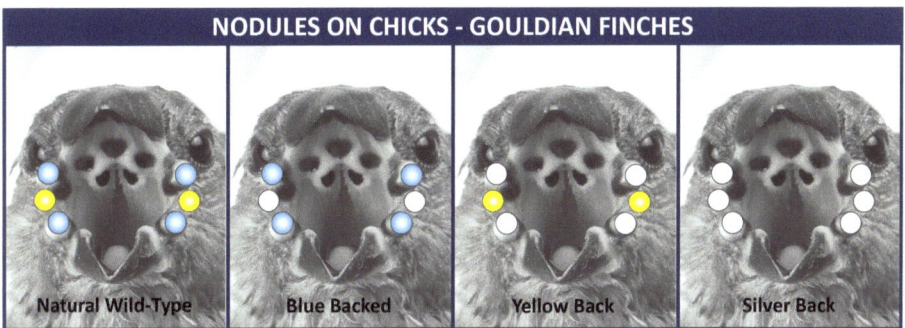
Nodule identification of nestlings

The Nodules on Nestlings

The nodules on baby Gouldian finches, known as papillae, are small, protruding structures situated inside their mouths.

Purpose

These nodules play a critical role during the early developmental stages of the chicks. They are brightly coloured and highly reflective, providing a vital visual cue for parent birds to accurately identify and locate the chicks' mouths when feeding in the dimness of the nest. This function is particularly crucial as the chicks are altricial and born blind. As the chicks mature, the nodules gradually diminish in prominence.

Differences Between Mutations

The colours of the nodules vary for different colour mutations, allowing these identifications to be made before adult feathers develop.

There are three nodules on each side of the nestlings' beaks:
- Blue – Yellow – Blue … Normal wild-type, Green Back
- Blue – White – Blue … Blue Back
- White – Yellow – White … Yellow Back (YB)
- White – White – White … Silver (Blue Back and YB)

Identifying Chicks in the Nest

Newly hatched Gouldian finch chicks are featherless with a distinctive skin tone which can help in identification. As they grow, subtle changes in their emerging feathers give clues about their potential adult for different colour mutations.

Being able to identify mutations in Gouldian chicks by their colour in the nest would be very informative

GOULDIAN CHICKS IN THE NEST			
	HATCHLING SKIN	FEATHERING COLOUR	FLIGHT FEATHER TIPS
Normal	Fleshy, pinkish-grey	Olive Green	Dark Grey
Dilute (SF YB)	Fleshy, pinkish-grey	Grey Green	Light Grey
Yellow Back (DF YB)	Pink	Yellow White	Very Light Grey
Blue	Pale Bluish Pink	Blue Grey	Grey
Pastel Blue (SF YB)	Pale Bluish Pink	Pale Blue Grey	Light Grey
Silver (DF YB)	Pale Bluish Pink	Grey White	White

EXAMPLES OF NESTLINGS

Normal Wild-Type

Pastel Blue

Blue

Yellow Back

for those who can become confident doing it. It is recognised that it is easiest for those clutches that also contain normal wild-type chicks for comparison.

Normal wild-type Gouldian chicks typically have a pinkish-grey skin tone that gradually develops into standard olive green coloration.

Chicks with the Blue mutation tend to have a noticeably lighter, almost bluish tint to their skin and as their feathers start to emerge they appear blue grey. Yellow Back chicks often have a more intense pink skin tone and their early feathers are yellowish-white. Pastel mutation chicks exhibit a softer, diluted version of the wild-type colours with a lighter skin tone and emerging feathers that have a more pale, greyish-pink colour.

Making these identifications is a case of experience being the best teacher. ("References Used in This Book" on page 236).

Using Foster Parents

Breeding Gouldian finches can be challenging due to their specific requirements and inconsistent parenting skills. Consequently, some bird breeders use alternative fostering methods, with Bengalese Finches often serving as surrogate parents.

Higher Survival Rates

Bengalese Finches exhibit strong parenting behaviours and commitment to caring for their offspring. They are less likely to leave their eggs or chicks unattended, which can result in a higher survival rate for fostered Gouldians.

Imprinting with Bengalese

Although Bengalese Finches can serve as surrogate parents for Gouldian chicks, they do not completely replicate the parenting behaviours of Gouldians. Gouldian chicks raised by Bengalese Finches may imprint on their foster parents. This could affect their social behaviour and mating preferences later in life, as they might face difficulties in recognizing and interacting with their own species.

An over-reliance on Bengalese Finches for fostering may result in Gouldian Finches becoming less capable of raising their own young. Although fostering can be helpful in specific situations, such as the death of the parents, it is not recommended as a standard practice.

Gouldians as Foster Parents

An alternative method for fostering is to use Gouldians that have previously shown good parenting skills. This approach helps to avoid any potential imprinting issues and is most commonly used when trying to establish new mutations among weaker birds.

The method for same-species fostering is simply to swap the eggs between two breeding pairs. However, it can be difficult to have a fostering pair at the same stage of the breeding cycle.

Candling Eggs to Check for Fertility

Candling is a method used to examine the internal contents of an egg, without causing any damage to it. This method is particularly valuable for bird breeders aiming to determine the fertility of their finch eggs.

Steps to Candle Finch Eggs

1. Prepare by having a small, bright pen-light. An LED model is ideal and one with a flexible neck is even better. Additionally, a clear plastic teaspoon can be useful. To achieve optimal contrast, conduct this activity in a darkened or dimly lit area; therefore, early evenings are often more suitable.

2. Care needs to be taken when handling delicate eggs with fingers. Hands should be thoroughly washed to prevent any potential contamination of the eggs.

3. Position the light source directly under the egg at its wider base, near the air bubble. Ensure that the light is focused on this area.

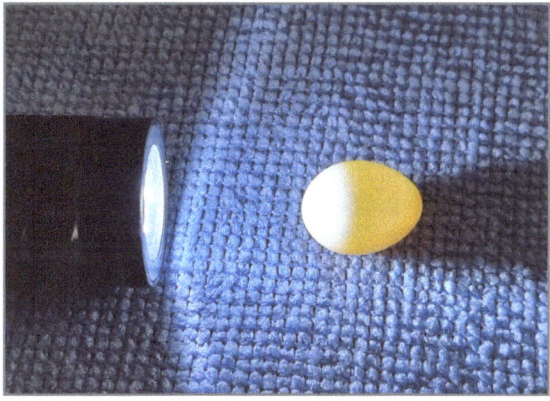

One simple method for candling an egg

4. Observe the pattern inside the egg and judge this appearance as described below.

Results When Candling Eggs

- Early Stage (1-4 days): The egg is mostly clear with a slight yolk shadow. Significant details are not usually observable during this period.
- Fertile Egg (after 5 days): Veins are visible along with a small dark spot representing the developing embryo. The network of veins becomes more noticeable over time.
- Non-fertile Egg (after 7 days): If the egg remains completely clear with no visible veins or embryo, it is likely to be infertile.

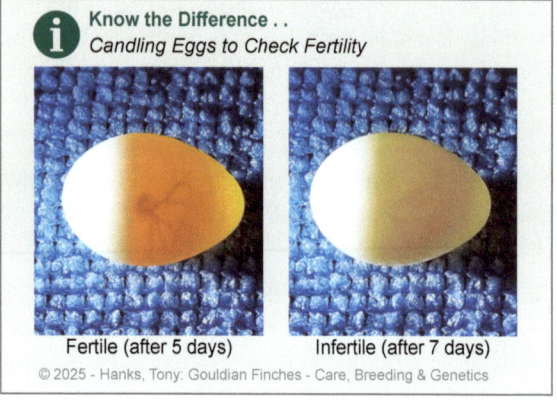

Successful candling requires careful handling of the eggs, no shaking and obviously no dropping of them.

Record Keeping for Breeding

Maintaining records when breeding Gouldian finches ensures accurate tracking of genetics, health history, and breeding outcomes. It also helps identify patterns or issues that may arise, allowing for adjustments in breeding strategies in subsequent seasons.

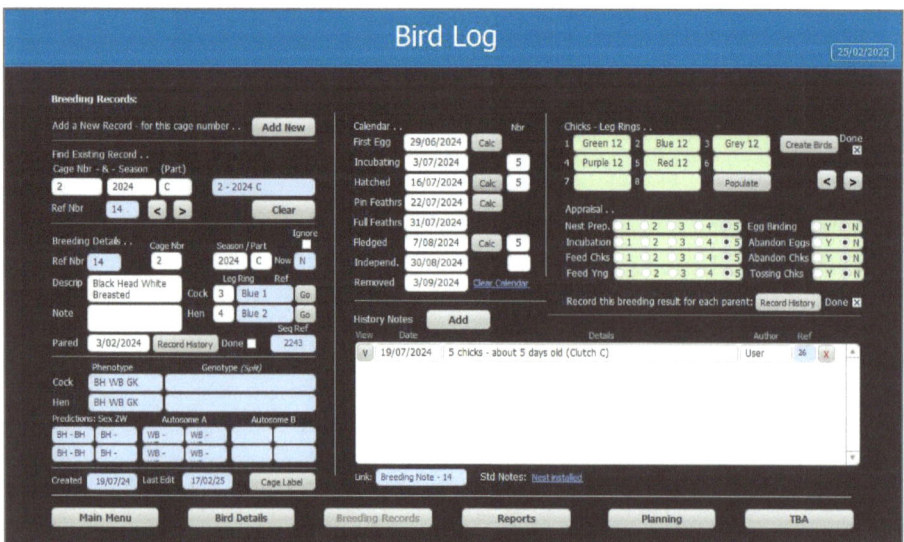

Example of a computer based Breeding Record

Breeding Log for Gouldian Finches

Ref Nbr	17	Cage Nbr	5	Season Yr	2025	Clutch Ref	B
Description	Dilute & Yellow Back			Note			

	Bird Ref	Leg Ring	Bird Description			
Cock	28	Green 7	RH Dilute (SF YB)		Date Paired	10/1/2025
Hen	36	Grey 4	RH Yellow Back		Date Nest	26/1/2025

Progeny Expected

- ☐ 100% Normal
- ☐ 50% Normal & 50% Split
- ☐ 100% Split
- ☑ 50% Split & 50% Mutation
- ☐ 100% Mutation
- ☐ 25% Normal, 50% Split & 25% Mutation
- ☐

	First Egg	Incubating	Hatched	Feathers	Fledged	Independent	Removed
Date		2/3/2025	18/3/2025		11/4/2025	5/5/2025	8/5/2025
Number	✕	4	4		4	4	4

Chicks	Leg Ring 1	Leg Ring 2	Leg Ring 3	Leg Ring 4	Leg Ring 5	Leg Ring 6	Leg Ring 7
	White 4	White 5	White 6	White 7			
Assessment	Nest Prep	Feed Chicks	Feed Young	Egg Binding	Leave Eggs	Leave Chicks	Chick Tossing
	5 /5	5 /5	5 /5	☐Y ☑N	☐Y ☑N	☐Y ☑N	☐Y ☑N

Date	General Notes, Pairings, Health, Medications, etc
26/1/2025	New balcony style nest
5/3/2025	Nesting material withdrawn

Example of a paper based Breeding Record

Record keeping ensures hens are not overburdened with too many clutches per season, protecting their health. It also confirms juvenile independence before separation, reducing stress for both young birds and the breeding pair.

Computer or Paper Records

The preferred methods for maintaining breeding records are either paper-based or digital. The illustration is an example of a screen from computer-based records.

A simpler record system utilizes paper, and a standard breeding record form for Gouldians can be found in the "Forms Library" section of this book. See page 203.

Line Breeding

Line breeding is a selective breeding method that is not generally recommended but sometimes used with Gouldians to enhance specific genetic traits. This technique involves mating closely related birds, such as siblings, cousins, or parent and offspring, to reinforce certain characteristics. While line breeding can result in improvements in attributes, it requires careful management to avoid potential issues such as inbreeding depression.

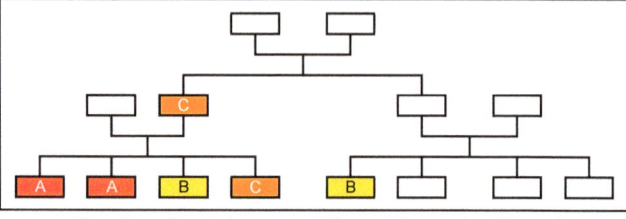

Line Breeding is between siblings (A), cousins (B) and parent to offspring (C)

Goals of Line Breeding

Line breeding has several objectives:
- Improving specific traits such as size, stature, or colour
- Establishing mutations
- Preserving rare traits

Risks of Line Breeding

Line breeding has several benefits, but Inbreeding Depression poses a significant risk, potentially resulting in reduced fertility, increased disease susceptibility and genetic defects. A natural outcome is decreased genetic diversity within a population. Ethical concerns can also arise if the practice leads to any adverse conditions for the birds involved.

There have been several examples of Inbreeding Depression during the establishment of the newer Gouldian mutations. It is often commented that these mutations are weaker, but this is more a result of line breeding than any true inherent deficiency of the mutation.

Managing Line Breeding

To reduce the risks of line breeding, it is recommended to select breeding pairs with minimal genetic weaknesses carefully. Also to incorporate Outcrossing into the breeding program, by introducing unrelated birds periodically to help improve genetic diversity. Additionally, regular health checks and monitoring are important to identify any potential issues.

Although line breeding presents opportunities to improve and maintain favourable traits, breeders must exercise care to reduce the associated risks and protect the health of the birds.

Mating Preferences Related to Head Colour

Assortative Mating

The head colour of Gouldian finches is important in mate selection and breeding success. Studies indicate that these birds exhibit "assortative mating," where they mainly select mates with similar head colours: Red-headed finches choose red-headed partners, black chooses black and yellow opts for yellow-headed mates.

Offspring Survival Rates & Gender Mixes

Researchers Sarah Pryke and Simon Griffith from Macquarie University in Sydney observed that pairs with matching head colours had higher offspring survival rates and the female offspring of mismatched pairs had lower survival rates. Additionally, mismatched pairs produced 82% male offspring, whereas pairs with the same head colour produced 46% males.

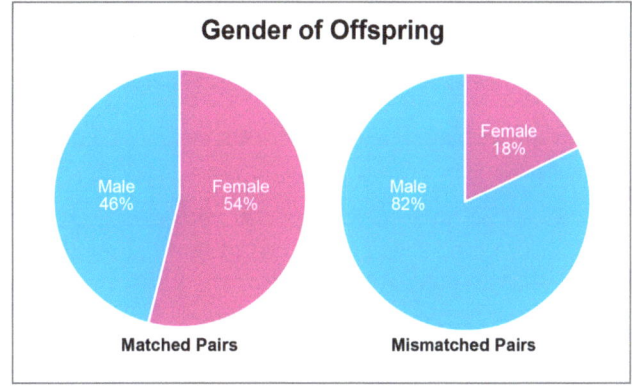

Analysis of the breeding records of the author has also confirmed the same higher offspring survival rates when pairs are matched for head colour. Pairs matched for head colour produce an average of 6 independent juveniles from up to two clutches per season, while this figure is closer to an average of 4 for mismatched pairs.

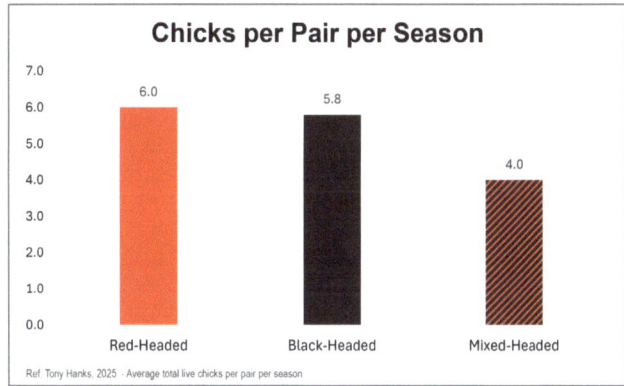

Behavioural Choices

Many other scientific studies have examined the relationship between head colour and mating preferences, such as Kang-Wook Kim and colleagues. Researchers have noted that red-headed Gouldian finches often exhibit higher reproductive success in competitive environments due to their assertive behaviour. In contrast, black-headed finches tend to perform better in calmer settings where their caution and conflict avoidance is a benefit.

Red-headed males, with their assertive and dominant behaviour, may be more attractive to red-headed females seeking strong and protective partners. On the other side, black-

AVERAGE GOULDIAN BREEDING RESULTS FOR SITTING HENS						
	NORMAL	MUTATIONS				
	Wild-Type	White Breasted	Yellow Back	Australian Yellow	Blue & Split	ALL
Hatched Rate ①	91%	86%	54%	71%	77%	74%
Fledged Rate ②	85%	61%	67%	58%	75%	65%
Offspring Survival Rate ③	77%	53%	36%	41%	58%	48%
Offspring per Pair ④	6.3	5.4	3.8	2.9	4.4	4.6

Ref: Tony Hanks, 2025. Results for nests with sitting hens, over 3 years.
① Chicks from fertile Eggs ② Fledglings from Chicks ③ Fledglings from fertile Eggs ④ Average total per season.

headed females may prefer the more cautious and reserved black-headed males, because they align better with their own temperament.

When Gouldian finches select mates with similar head colours, they enhance the probability of producing healthy offspring with desirable traits, genetic compatibility, and optimised reproductive success.

Dominant Doesn't Dominate

This also explains why the dominant Red-Headed gene has not completely eliminated the recessive Black-Headed version: A balance has been achieved because each head colour has its' own advantages in behaviour, together with there being a mating preference and more balanced sexes in the offspring numbers when matched head colours mate. (See "References Used in This Book" on page 236).

Breeding Results for Mutations

When people start breeding Gouldian finch mutations, they often observe that these finches are more challenging to breed compared to the normal wild-types. Although some breeders suggest that they do not need special treatment, there are noticeable differences in breeding outcomes, as illustrated in the author's table of results.

Average Breeding Results

Over a three-year period, the controlled breeding of various mutations yielded an average of 4.6 offspring per pair per season. This outcome is 27% lower than the average of 6.3 offspring observed in normal wild-type pairs

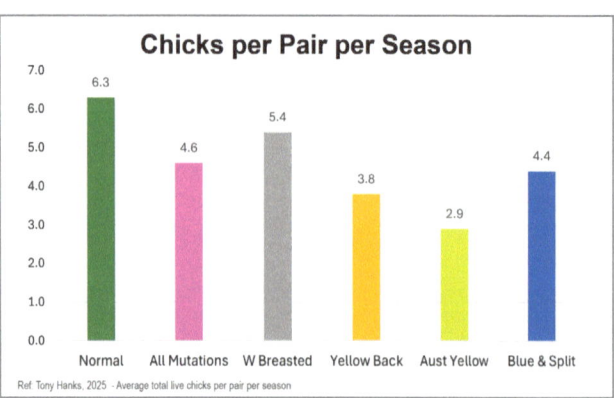

during the same seasons and conditions.

Fewer chicks hatched from eggs, and fewer nestlings survived to become fledglings. Understanding these factors can help explain the lower rate of breeding success.

Reasons for Reduced Breeding of Mutations

1. As previously discussed, the ability to perceive the colours of a potential mate is an important influence on partner selection. This is also why full spectrum lighting is crucial. Therefore, any mutations that diminish these visual signals will inevitably impact this process.
2. With a preference for certain head colours in mate selection, any mutation that dilutes or eliminates these colours will naturally have a consequent affect upon this process.
3. Colouration is a vital signal in the choice of mate. Richer more intense colours indicate more desirable traits, so mutations that minimise these signs can be less desirable.
4. Some mutations may have occurred through line breeding and inbreeding, which can result in genetic weaknesses that affect outcomes.
5. Mutation breeding programs may have used foster parents to achieve better results. This practice causes imprinting, making birds raised this way poorer parents.
6. Another potential factor is the age of the birds and the patience of the owner. Maturity can positively influence Gouldian breeding success, and it is important for the birds to have sufficient time for pair bonding before the mating season.

Selling Offspring

The result of almost all successful Gouldian breeding programs will be excess birds. So after planning the breeding pairs for the following year, it will be time for the remaining young birds to find new homes.

Readiness to Sell or Move

Gouldian finches should only be moved to a new home after they have completed their first annual moult. The moult itself is a stressful time for all birds and there is also the possible issue of "Cage Fright", where a badly handled move can cause stress, even leading to death. Another reason to wait is so that the sex of each bird becomes known - for the current owner to plan what to sell and for the owner to identify the birds.

Friends and Bird Club Members

It is always a good idea to share birds between people you know as friends or members of your Bird Club. There is a trust that gives confidence and the ability for them to follow up if they realise later that they have questions.

Bird Expos

These opportunities for selling birds are operated by Bird Clubs, generally once or twice a year. Potential buyers can meet sellers and ask questions, prices are competitive and all sales can usually be completed in a few hours.

Bird Dealers or Local Pet Stores

Another good option are registered Bird Dealers. The price will generally be lower because the dealer needs a margin when re-selling. However, this is a very convenient method for large numbers of birds to be sold in a single transaction. Bird Dealers are usually also long-term bird enthusiasts, so they have a potential wealth of experience.

Community pet stores are usually more convenient, but they are a less popular option for selling Gouldians. This is because they are general pet shop where birds are a minor part of that business. A large proportion of buyers here are likely to buy a bird as a "pet" in this environment, which may mean that the Gouldian is kept alone and in an unsuitable cage.

Online Marketplaces

With the rise of social media there are now many options for selling birds online. Prices can be attractive, but a downside is that people you don't know will be coming to your home to consider their purchase.

Information to Accompany a Sale

Any serious purchaser of Gouldian finches will want to know the background information for each bird. This includes the genetics of each bird and any information about split genes that do not appear in the phenotype. A description of the parents is also helpful, along with any notes about diet and health.

A Sale Transfer Form is included in the "Forms Library" on page 203.

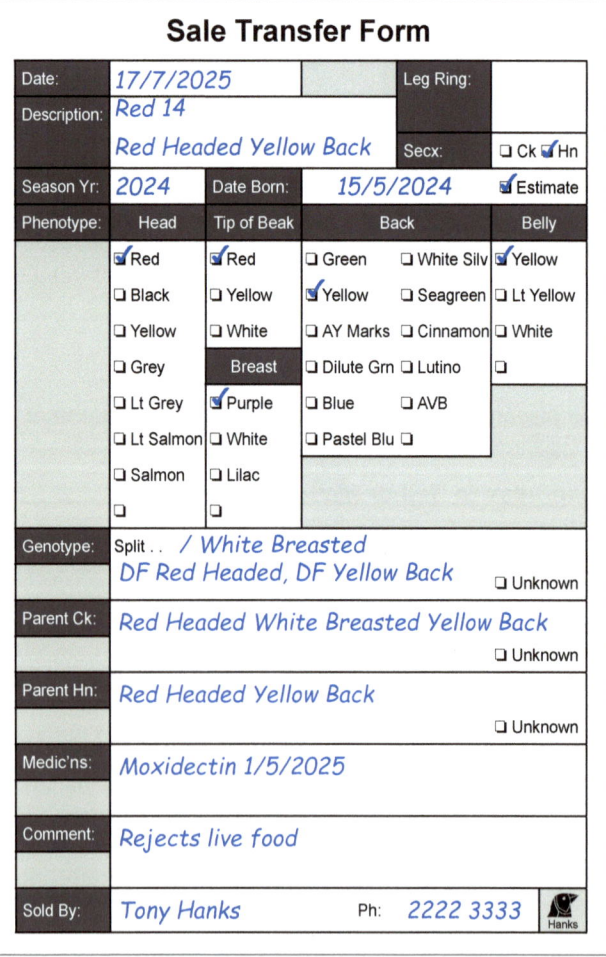

Example of a completed sale transfer form

7

GENETIC CONSIDERATIONS

Many people believe that they do not really need to understand genetics in order to breed Gouldian finches, while others report this as one of the most satisfying aspects. The reality is that understanding the genetics of birds is crucial for several reasons.

Why Understand Genetics

By studying genetics, breeders can prioritise the health and well-being of the birds, while also avoiding genetic disorders. Knowledge of this also allows the selection of breeding birds for desirable traits.

Maintaining genetic diversity within a breeding population is essential for the long-term adaptability and resilience of the species. Inbreeding needs to be avoided to maintain a varied gene pool and resist diseases.

Genetics also play a role in the appearance and behavioural traits of birds, such as head, chest and back colours, temperament, nesting habits and song patterns. Understanding these genetic influences can aid in breeding birds that are better suited for captivity or exhibition.

In summary, the study of genetics in bird breeding is not just about creating visually appealing birds but also about ensuring their health, diversity and ability to thrive. It provides a scientific foundation for making informed breeding decisions that benefit both the birds and their carers.

Mendelian Inheritance

The genetics of Gouldian finches follows the classical discoveries Gregor Mendel and Mendelian Inheritance. These principles explain how traits are passed from one generation to the next. Traits are determined by alleles, which are different forms of a gene, and alleles can be dominant or recessive. Dominant alleles express their traits in the appearance or phenotype even if only one copy is present, while recessive alleles require two copies for them to be expressed.

Individuals can be homozygous or heterozygous for a trait. Those who are homozygous have two identical alleles for a trait (example AA or aa), while those who are heterozygous have two different alleles (example Aa). Then in heterozygous individuals, the dominant allele masks the effect of the recessive allele.

Sex Chromosomes of Birds

Many people may have learned about sex chromosomes during their science education at school or through subsequent readings. In mammals, including humans, males possess XY chromosomes while females possess XX chromosomes. This differs in birds due to divergent evolutionary pathways, because birds and mammals did not evolve from one another.

In birds, it is the males that have identical ZZ sex chromosomes, whereas females have differing ZW sex chromosomes. The sex chromosomes of males are said to homologous and those of females are heterologous.

Genetics of Each Bird

Before discussing the genetics of breeding, it is necessary to understand the genetics of each individual bird.

The outward appearance of each bird is called its' phenotype, while the characteristics carried on the chromosomes are called the genotype. Gouldian finches have 78 chromosomes (39 pairs), two of which are sex chromosomes.

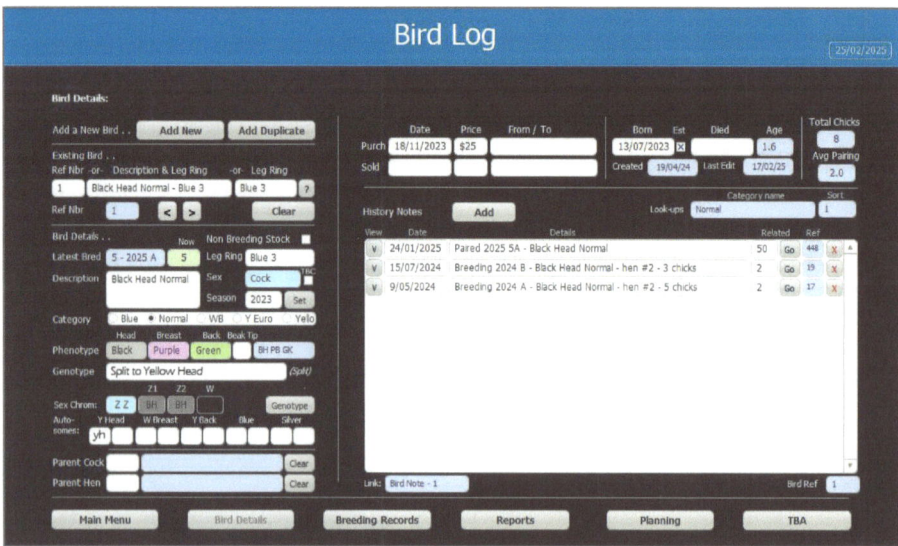

Example of a computer based Bird Record

Computer or Paper Records

Just as it important to keep accurate Breeding Records, these will be based upon having individual Bird Records. Once again these can be either paper-based or digital.

The image is an example of a screen from computer-based records and a simpler record system using paper is included in the "Forms Library" section of this book. See page 203.

Phenotype Summaries

Many breeders of Gouldian finches use a summary for the outward appearance or phenotype. However, only good record keeping give information about the genotype.

Phenotype summaries (appearance) use the following abbreviations:

					Legend							
H	Head	L	Belly	B	Black	G	Green	LS	Light Salmon	R	Red
B	Breast	T	Tail	B	Blue	G	Grey	LY	Light Yellow	S	Salmon
K	Back	TB	Tip of Beak	BG	Blue-Green	LB	Light Blue	M	Mauve	W	White
C	Chin			D	Dilute	LG	Light Grey	OG	Olive Green	Y	Yellow
N	Nape			DB	Dark Blue	LR	Light Red	P	Purple		

For example, noting that these abbreviations refer to the appearance - not the genetics:

- Red-headed normal   RH PB GK
- Black-headed normal   BH PB GK
- Black-headed white breasted   BH WB GK
- Red-headed Australian Yellow RH WB YK
- Black-headed blue backed   BH PB BK

Understanding genetics adds a lot of enjoyment to aviculture

Examples of common phenotype summaries used in bird records . .

PHENOTYPE	Description	Red Headed Normal			
	Sex	Cock			
	Abbrev.	RH	PB	GK	RTB

PHENOTYPE	Description	Black Headed Normal			
	Sex	Hen			
	Abbrev.	BH	PB	GK	RTB

PHENOTYPE	Description	Yellow Headed Normal			
	Sex	Cock			
	Abbrev.	YH	PB	GK	YTB

PHENOTYPE	Description	Black Headed split to DF Yellow			
	Sex	Hen			
	Abbrev.	BH	PB	GK	YTB

PHENOTYPE	Description	Red Headed White Breasted			
	Sex	Cock			
	Abbrev.	RH	WB	GK	RTB

PHENOTYPE	Description	Black Headed Blue Backed			
	Sex	Hen			
	Abbrev.	BH	PB	BK	WTB

PHENOTYPE	Description	Red Headed Australian Yellow (AY)			
	Sex	Cock			
	Abbrev.	RH	WB	YK	RTB

PHENOTYPE	Description	Red Headed Yellow Back (YB)			
	Sex	Hen			
	Abbrev.	RH	PB	YK	RTB

Genotype Summaries

A standard report format can also be used to summarize the genetic characteristics of each bird. Some traits are carried on the sex chromosomes, while others are autosomal. Additionally, some traits are dominant, while others are recessive, requiring two copies of the same gene in order for the trait to be expressed in the phenotype (outward appearance). See "Quick Reference Table - Genetics of Gouldian Finches" on page 139, later in this chapter.

Single Factor or Double Factor

Genetic inheritance is based upon pairs of genes. When two genes are present for the same trait that is said to be "Double Factor", while just one gene for that trait is "Single Factor". Of course, if it is a dominant gene then having just one (Single Factor) will be sufficient for the trait to appear in the phenotype (appearance).

TIP The genetics term "Double-Factor" or "DF" means the same as "Homozygous". It is the same case for "Single-Factor", "SF" and "Heterozygous".

Genotype Examples

The terminology for the genotypes can be used to show the various traits as they appear on the chromosomes. By referring to these symbols the bird can be identified, progeny predicted and breeding planned.

Breeding for Health

When selecting pairs of birds for breeding it is important to prioritise health over appearance. It can be tempting to select birds for breeding only on the basis of expected outcomes, but this effort will be wasted if the offspring are weak or carry genetic disorders.

Achieving this balance with Gouldian finches is simply adhering to responsible breeding practices.

Colour Mutations

Exploring the different colour mutations in Gouldian finches can be both fascinating and important for breeders aiming to achieve specific traits.

The normal wild-type Gouldian finches have strikingly colourful appearances, but there is also an array of colour mutations that captivate breeders. These mutations, which can affect both the body and head colours, are the result of specific genetic variations and can range from subtle shifts to dramatic transformations.

Genotype Conventions

Upper case letters are used dominant genes and lower case is used for those

GENOTYPE EXAMPLES	
Symbols	Description
<Upper Case>	Dominant gene
<Lower Case>	Recessive gene
(sex chromosome symbol)	Sex chromosome
(autosome symbol)	Autosome
Z:R Z:R	Red-Headed cock
Z:R W	Red-Headed hen
Z:b Z:b	Black-Headed cock
Z:b W	Black-Headed hen
Z:R / yh Z:R / yh	Yellow-Headed cock
Z:R / yh W	Yellow-Headed hen
Z:R Z:b	Red-Headed cock (split Black)
Z:R Z:R / yh	Red-Headed cock (split Yellow)
Z:R / yh W	Red-Headed hen (split Yellow)
Z:b / yh Z:b	Black-Headed cock (split Yellow SF)
Z:b / yh Z:b / yh	Black-Headed cock (split Yellow DF)
Z:b / yh W	Black-Headed hen (split Yellow SF)
Z:b / yh W / yh	Black-Headed hen (split Yellow DF)

that are recessive (needing two copies to be seen in the appearance or "phenotype"). For the symbols, rounded corners indicate sex linked and bevelled corners are used for the autosomes.

Head Colours

In terms of head colour, there are three primary wild-type variations: red-headed, black-headed and yellow-headed (also called orange-headed). The red-head trait is sex-linked dominant to the black-headed form, whereas the yellow-head mutation is autosomal recessive and appears only when the bird is otherwise red-headed. See "Quick Reference Table - Genetics of Gouldian Finches" on page 139.

Other head colours like salmon and grey are also possible in the mutations of Gouldian finches. These are actually all variations in the appearance of red, black and yellow heads and they are discussed later under the heading of "Head Colours in Mutations" on page 132.

HEAD COLOURS				
Trait	Abbrev	Inheritance	Sex Linked	Autosomal
Red	RH	Dominant sex-linked	Z:R Z:R	
			Z:R Z:b	
			Z:R W	
Black	BH	Recessive sex-linked	Z:b Z:b	
			Z:b W	
Yellow	YH	Recessive autosomal		yh yh

Breast Colours

The breast colour is purple in normal Gouldian finches, while white-breasted and lilac-breasted mutations have also been developed. See "White Breasted" in the "Quick Reference Table - Genetics of Gouldian Finches" on page 140.

BREAST COLOURS				
Trait	Abbrev	Inheritance	Sex Linked	Autosomal
Purple	PB	Dominant autosomal		PB PB
				PB lb
				PB wb
Lilac	LB	Recessive autosomal but dominant to White		lb lb
				lb wb
White	WB	Recessive autosomal		wb wb

→ **Applicable to Most** - Note that these breast colours are possible in almost all of the Gouldian finch mutations.

→ **Epistatic Exceptions** - However, notable exceptions are the Australian Yellow and the Lutino, where the white breast colour is a feature of the pigment changes in that mutation - rather than the specific genes listed above for breast colour. This is called an "Epistatic Mutation" where one gene can mask the expression of another gene.

→ **Possibly Masked in the Phenotype** - Australian Yellow and Lutino birds can carry genes for breast colour in their genotype, but the breast colour is always white in their phenotype.

White Breasted is a common Gouldian mutation

Back Colours

The most common body colour mutations are from green to yellow, and blue. The wild-type Gouldian finch displays a green body, but through selective breeding, yellow-backed and blue-backed finches have been developed. These variations arise from recessive genes, meaning that the bird must inherit two copies of the mutation for the trait to be visible in the phenotype. The exception is the Yellow Back variation

Genetic Considerations

(European) where the inheritance is sex-linked, so a female displays the trait when it is only on one chromosome – since she only has one Z chromosome.

Blue-Backed Mutation

Blue backed Gouldians display the blue colour because they are unable to express the normal intensity of yellow colouration in their feathers, so the green back appears blue, the yellow belly is white and a red head becomes a light salmon colour. The mode of inheritance is recessive autosomal.

The gene for Blue is recessive autosomal, as discussed later. This means that it needs to be double-factor in the genotype for it to be expressed in the phenotype (appearance). For more details see "Blue Backed" in the "Quick Reference Table - Genetics of Gouldian Finches" on page 141.

Blue Back mutations

When breeding Blue Gouldian finches a pairing of Blue x Blue is not recommended due to probable further weakening of the Blue gene pool. Instead, the recommended pairing is Blue x Split Blue, or Blue x Normal to produce all Split Blues and strengthen the bloodlines.

As fledglings the Blue mutation can be distinguished by the grey-blue colour of these juveniles.

Yellow-Backed Mutations

Yellow backed Gouldian finches were developed independently of one another in Europe and Australia. Although both of these mutations have yellow backs, they have very different genotypes.

Note: Over time and in different countries there has been considerable confusion in the naming of these two versions, so the alternative names are included in case they might be useful when obtaining information from other sources.

Yellow Back

(Also known as "European Yellow-Backed", "Pastel" and "European Pastel"). The mode of inheritance of this trait is dominant sex-linked. Since it occurs alongside the gene for head colour it is referred to as "co-dominant".

Yellow Back mutation (DF)

This means that cocks can be single factor or double factor, with single factor appearing like a dilute version of a normal wild-type bird. Single factor Yellow Back Gouldian cocks are called "Dilute", "Euro Dominant Dilute", or "Pastel Green". Since hens have only one Z chromosome they cannot occur in this Dilute form. See "Dilute" on page 140.

Dilute (SF Yellow Back) mutation

The double factor version shows any normally black areas as white, while the single factor shows them as grey. Hens are only single factor because they have only one Z chromosome, but they appear similar in colour to the double factor cocks. See "Yellow Back" in the "Quick Reference Table - Genetics of Gouldian Finches" on page 140.

TIP It is not recommended to pair the Yellow Back and Australian Yellow mutations. The mode of inheritance is so different that it would be easy to lose track of the genotypes in the offspring.

Australian Yellow

(Also known as "Australian Yellow-Backed" or "Australian White Breasted Yellow"). In this case the inheritance method is autosomal recessive. Birds showing the trait can have a red or yellow head, however the black-headed variety appears as a silver-grey colour since the bird is devoid of melanin.

These birds always have a white breast, but this appearance is unrelated to the "wb" gene for White-Breasted. Instead the white breast is an epistatic part of the Australian Yellow mutation, but each bird may still carry unexpressed genes for breast colour in their genotype.

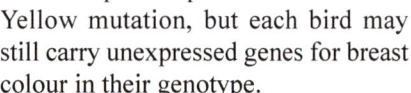

They also have varying degrees of brown-green markings running through the yellow back, called "foul feathers". for more details see "Australian Yellow" in the "Quick Reference Table - Genetics of Gouldian Finches" on page 141.

Australian Yellow mutation

The Australian Yellow mutation is also a special exception in that split birds, carrying just one copy of this recessive gene, will sometimes show a sign in their phenotype (appearance). Often they have

Normal split to Australian Yellow

some unexpected white feathers on the otherwise normal looking bird. This is a good indication of a split Australian Yellow, but it can only be confirmed in breeding trials.

Identifying the Yellow Backed Version

Due to the distinct genetic differences between the Yellow Back and the Australian Yellow, accurate identification of each bird is crucial when making breeding selections. The primary distinction is that the Australian Yellow exhibits a lighter shade of yellow on its back and possesses some green markings amongst the yellow feathers. As a distinction, the Yellow Back (European) birds have a very clean yellow back.

The Australian Yellow has a white breast, but the Yellow Back can have a purple or lilac breast in double factor males and a purple, lilac or white breast in single factor males and females.

It is important not to confuse these two different mutations

Australian Recessive Dilute

This is also known as "Australian Dilute" or "Lime". Birds of this mutation are sometimes confused with the Dilute single-factor version of the Yellow Back, but it is genetically completely different. While the Dilute version of Yellow Back is a dominant sex-linked gene, the inheritance for the Australian Recessive Dilute is autosomal recessive. It is therefore possible for this mutation to occur for both cocks and hens.

There is also a visible difference, with the Australian having a back colour that is a yellow-green, while the Dilute single-factor Yellow Back is a light green. For more details see "Australian Recessive Dilute" in the "Quick Reference Table - Genetics of Gouldian Finches" on page 143.

These are also two different mutations

Cinnamon

This inheritance method is sex-linked recessive. Birds showing this trait have paler bodies with reduced carotenoid pigmentation by approximately 50%. Effectively the Eumelanin Black of the Normal bird is changed to Brown. In some publications this mutation is also referred to as Bruno (meaning "brown"). Like most other mutations,

Cinnamon can also be combined with White Breasted. For more genetic details see page 143.

Seagreen

This inheritance method is also sex-linked recessive. For details see "Seagreen" in the "Quick Reference Table - Genetics of Gouldian Finches" on page 144.

Fallow

(Also known as "European Fallow"). This inheritance method is autosomal recessive and these birds display softer, lighter hues - primarily impacting melanin-based colours. Black appears more brown; green is more olive; and reds and yellows are more pastel. There is also a slight redness to the eyes because of reduced melanin in the iris. See "Fallow" in the "Quick Reference Table - Genetics of Gouldian Finches" on page 144.

Lutino

Birds with the Lutino mutation appear as yellow and white, with the inheritance method being sex-linked recessive. The white breast colour is again epistatic, as a result of the Lutino mutation masking any alternative breast colour in the genotype. See page 144.

Single vs Double Factor Mutations

As shown earlier, other variations and colour combinations are possible when chromosomes are single or double factor, or when different traits are combined in the same bird.

The following are examples of recognised single and double factors:

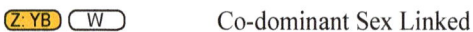

Yellow Back Cocks . . .   YB
- Single Factor YB - Dilute   (Z: YB) ( )   Co-dominant Sex Linked
- Double Factor YB   (Z: YB) (Z: YB)   Co-dominant Sex Linked

Yellow Back Hens . . .
- Single Factor YB   (Z: YB) (W)   Co-dominant Sex Linked

Combined Mutations

These are Gouldians where two or more mutations exist at the same time, but where they affect different parts of the bird so they each appear the same as expected. For example:

White-Breasted Yellow Back

As the name suggests, these birds are a combination of the autosomal White-Breasted and the sex-linked Yellow Back mutations. Note that two different genotype variations produce almost the same

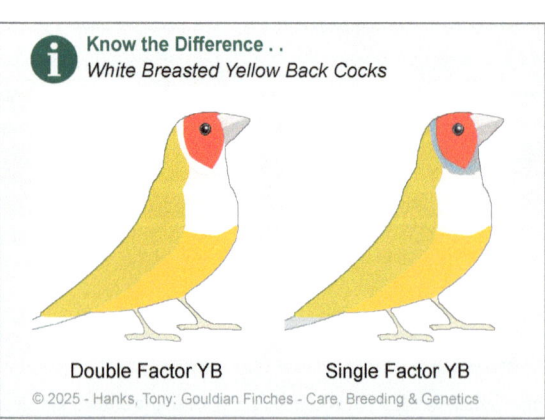

Know the Difference . .
White Breasted Yellow Back Cocks

Double Factor YB   Single Factor YB

© 2025 - Hanks, Tony: Gouldian Finches - Care, Breeding & Genetics

In this case the colour on the chin & nape reveals the genotype

phenotype in cocks. The distinction is that the chin, nape and tail are light grey in the single-factor version. See page 140.

White Breasted Yellow Back Cocks . . .   *WB YB*
- Single Factor Yellow Back   (Z: YB) ( ) (wb) (wb)
- Double Factor Yellow Back   (Z: YB) (Z: YB) (wb) (wb)

White Breasted Yellow Back Hens . . .
- Single Factor Yellow Back   (Z: YB) (W) (wb) (wb)

White-Breasted Blue

In this case the mode of inheritance is all autosomal for both the White-Breasted and the Blue Back mutations. For more details see "Blue Backed White Breasted" in the "Quick Reference Table - Genetics of Gouldian Finches" on page 141.

The full list is in the Quick Reference Table, but these following examples illustrate some combinations of inherited traits:

- Black Headed Blue Cock   (Z: b) (Z: b) (bk) (bk)
- Black Headed White Breast Blue Cock   (Z: b) (Z: b) (wb) (wb) (bk) (bk)
- Red Headed Yellow Back Hen   (Z: R) (W) (Z: YB) (W)
- Yellow Headed Yellow Back Hen   (Z: R) (W) (Z: YB) (W) (yh) (yh)

Secondary Mutations

These secondary mutations occur when two or more mutations are combined in the same bird, acting upon the same areas to produce a result that is different from the phenotype of either mutation alone. The most recognised of these are as follows:

Pastel Blue

This is another example where double vs single factor makes a large difference to the phenotype (appearance).

While the Pastel Blue mutation includes Single-Factor Yellow Back genes, changing those to Double-Factor results in Silver. See page 142.

Note that some texts have confusingly used the words "Pastel Body" or "SF Pastel" to describe the Dilute mutation; and "DF Pastel" to describe the Yellow Back.

Pastel Blue Cocks . . .   *PK*
- DF Blue and SF Yellow Back   (Z: YB) ( ) (bk) (bk)

(Pastel Blue is not possible in hens)

Silver

For more details see "Silver" in the "Quick Reference Table - Genetics of Gouldian Finches" on page 142.

Silver Back Cocks . . .   *SK*
- DF Blue and DF Yellow Back   (Z: YB) (Z: YB) (bk) (bk)

Silver Back Hens . . .
- DF Blue and SF Yellow Back   (Z: YB) (W) (bk) (bk)

Pastel Blue mutation

White Breasted Silver mutation

White-Breasted Silver

In this example there are three mutations combined - White Breasted, Blue Backed and Yellow Back. For details see "Silver White Breasted" in the "Quick Reference Table - Genetics of Gouldian Finches" on page 142.

Silver Back White Breasted Cocks . . .   *WB SK*
- DF Blue and DF Yellow Back and DF White Breasted
  (Z: YB) (Z: YB) (bk) (bk) (wb) (wb)
- DF Blue, SF Yellow Back and DF White Breasted
  (Z: YB) ( ) (bk) (bk) (wb) (wb)

Silver Back White Breasted Hens . . .
- DF Blue and SF Yellow Back and DF White Breasted
  (Z: YB) (W) (bk) (bk) (wb) (wb)

Australian Recessive Dilute Blue

This attractive secondary mutation is a combination of these two autosomal recessives. For details see "Australian Recessive Dilute Blue" in the "Quick Reference Table - Genetics of Gouldian Finches" on page 143.

Ivory

The phenotype known as Ivory is a combination of the autosomal Blue and the sex-linked Cinnamon. For details see "Ivory" in the "Quick Reference Table - Genetics of Gouldian Finches" on page 145.

Satine

This secondary mutation is also called "Satinet" and it results from the combination of two recessive sex-linked mutations - Cinnamon and Lutino. For more details see "Satine" in the "Quick Reference Table - Genetics of Gouldian Finches" on page 145.

Australian Variegated Blue

This secondary mutation is a combination of two autosomal mutations - the recessive Australian Yellow and the recessive Blue. See the genetic details on page 145.

The breast is white and the back colour varies from a light to dark grey-blue melanin

Genetic Considerations

Trait	Abbrev	Inheritance	Sex Linked	Autosomal
BACK COLOURS - SUMMARY of INHERITANCE				
Green	GK	Dominant autosomal		*Normal*
Blue	BK	Recessive autosomal		bk bk
Yellow YB	YB	Co-dom'nt sex-linked	(Z: YB) (Z: YB) or (Z: YB) (W)	
Dilute YB - SF Cock	DK	Co-dom'nt sex-linked SF	(Z: YB) ( )	
Yellow AY	AY	Recessive autosomal		ay ay
Aust Recessive Dilute	AD	Recessive autosomal		ad ad
Cinnamon	CK	Recessive sex-linked	(Z: cn) (Z: cn) or (Z: cn) (W)	
Seagreen	SGK	Recessive sex-linked	(Z: sg) (Z: sg) or (Z: sg) (W)	
Fallow	FK	Recessive autosomal		fa fa
Lutino	LUT	Recessive sex-linked	(Z: ino) (Z: ino) or (Z: ino) (W)	
Wh B - Yellow Back	WB YB	(Y. Back & Wh Breast)	(Z: YB) (Z: YB) or (Z: YB) (W)	wb wb
Pastel Blue - SF Cock	PK	(Y. Back SF & Blue)	(Z: YB) ( )	bk bk
Silver	SK	(Y. Back DF & Blue)	(Z: YB) (Z: YB) or (Z: YB) (W)	bk bk
Wh B - Silver	WB SK	(Y. Back DF & Blue & Wh B)	(Z: YB) (Z: YB) or (Z: YB) (W)	bk bk / wb wb
Wh B - Silver - SF Cock	WB SK	(Y. Back SF & Blue & Wh B)	(Z: YB) ( )	bk bk / wb wb
Aust Recess Dilute Blue	ADB	(Aust Recessive Dilute & Blue)		ad ad / bk bk
Ivory	IK	(Cinnamon & Blue)	(Z: cn) (Z: cn) or (Z: cn) (W)	bk bk
Satine	SAT	(Cinnamon & Lutino)	(Z: cn) (Z: ino)	
AVB	AVB	(Aust Yellow & Blue)		ay ay / bk bk

pattern against a white ground colour. Hens tend to be greyer and cocks bluer, but this is not sufficient for reliably sexing these birds.

It can be helpful to note that normal birds split for AVB often show a white spot under the beak. Sometimes the Australian Variegated Blue is also referred to as Australian Yellow Blue (AYB).

Combinations of mutations and secondary mutations are all listed in the table titled "Quick Reference Table - Genetics of Gouldian Finches" on page 139, however these are examples:

- Black Headed Pastel Blue Cock   (Z: b) (Z: b) (Z: YB) (W) bk bk
- Red Headed Silver White Breast Hen   (Z: R) (W) (Z: YB) (W) (wb) (wb) bk
  bk

See the modes of inheritance summarised for all "Back Colours" on page 131.

Alongside these primary mutations, breeders have also identified combinations of these traits, producing birds with unique and diverse appearances. For instance, red-headed, blue-backed Gouldian finch is a particularly striking combination that highlights the potential of genetic diversity within this species.

Understanding all of these colour mutations is essential for breeders who aim to produce specific appearances in their birds. However, it is also crucial to approach breeding with a focus on the health and well-being of the birds, ensuring that genetic diversity is maintained and that the offspring are robust and free from genetic disorders.

Head Colours in Mutations

Head colours in mutations that are not the natural wild-type Gouldians can appear differently than what the genotype might have predicted.

Changed Head Colour Appearances

This occurs because some mutations result in a phenotype (appearance) due to the inability to produce a specific colour in the feathers.

For example, the Yellow Back mutation is that colour because it is unable to produce black, so the Black Head trait appears as light grey:

A complete list of these head colour variations in the phenotypes for different genetic combinations are listed in the table titled "Quick Reference Table - Head Colours in Gouldian Mutations" on page 133.

Dominant & Recessive Traits

Dominant Genes

As the name implies, dominant genes are those that express their traits when only one copy of the gene is present (heterozygous). This means that if a bird inherits a dominant gene for a specific trait from one parent, that trait will be visible in the bird's appearance. For example, in Gouldian finches, the red-head trait is determined by a dominant gene. Therefore, a finch only needs to inherit one red-head gene from either parent to display a red head.

Recessive Genes

By contrast, recessive genes require two copies (homozygous) to express the corresponding trait. This is also referred to as being "double factor" (DF). If a bird inherits only one copy of a recessive gene, the presence of a dominant gene will mask the trait. However, the bird will still carry the recessive gene and can pass it on to its' own offspring, a condition often referred to as being "split" or "single factor" (SF) for that characteristic. For example, the blue-backed mutation in Gouldian finches is a recessive trait, requiring the finch to inherit the blue-backed gene from both parents to display a blue back.

Understanding dominant and recessive genes is important for breeders to predict the probability of traits appearing in offspring. By selecting breeding pairs based on their genetic makeup, breeders can achieve specific traits and maintain the health and genetic diversity of the finch population.

Genetic Considerations

Quick Reference Table - Head Colours in Gouldian Mutations

Natural or Mutation	Purple-Breasted			White-Breasted		
	Red-Head	Black-Head	Yellow-Head	Red-Head	Black-Head	Yellow-Head
Natural Wild-type	Red	Black	Yellow	Red	Black	Yellow
Yellow Back (YB DF)	Red	Lt Grey	Yellow	Red	Lt Grey	Yellow
Dilute or SF Yellow Back (YB SF)	Red	Grey	Yellow	Red	Lt Grey	Yellow
Aust. Yellow (ay DF)	(NA)	(NA)	(NA)	Red	Very Lt Grey	Yellow
Blue (bk DF)	Salmon	Black	Salmon	Lt Salmon	Black	Lt Salmon
Pastel Blue (bk DF & YB SF)	Lt Salmon	Grey	Lt Salmon	(NA)	(NA)	(NA)
Silver (bk DF & YB DF -or- bk DF, YB SF & WB)	Lt Salmon	Very Lt Grey	Lt Salmon	Lt Salmon	Very Lt Grey	Lt Salmon
Australian Recessive Dilute (ad DF)	Red	Lt Grey	Yellow	Red	Lt Grey	Yellow
Cinnamon (cn DF)	Red	Grey	Yellow	Red	Grey	Yellow
Seagreen (sg DF)	Lt Red	Black	Lt Yellow	Lt Red	Black	Lt Yellow
Aust. Variegated Blue (AVB) (ay DF & bk DF)	(NA)	(NA)	(NA)	Lt Salmon	Grey	Lt salmon

© 2025 - Hanks, Tony

Gouldian Finches - Care, Breeding & Genetics

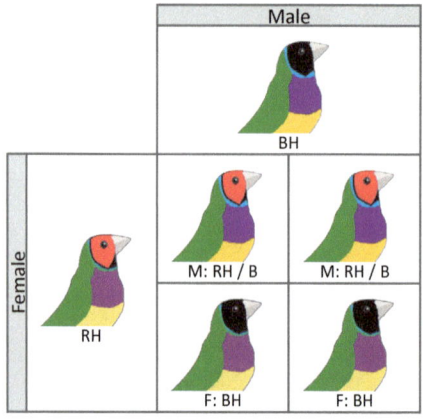

Genetic Inheritance in Gouldians

The modes of inheritance for Gouldian Finches can be sex-linked or autosomal, consistent with the classical genetics as described by Mendel. This is an example to illustrate the expected results of a common mating. Note that while the cock is black-headed and the hen is red-headed, all of the possible offspring are the opposite. Understanding why this happens is the power of genetics.

Sex-Linked Inheritance

Sex-linked inheritance refers to the transmission of traits that are determined by genes that are located on the sex chromosomes.

Z & W Chromosomes

In birds, the sex chromosomes are designated as Z and W, with males possessing two Z chromosomes (ZZ) and females having one Z and one W chromosome (ZW).

Sex-linked traits are controlled by genes on the Z chromosome:

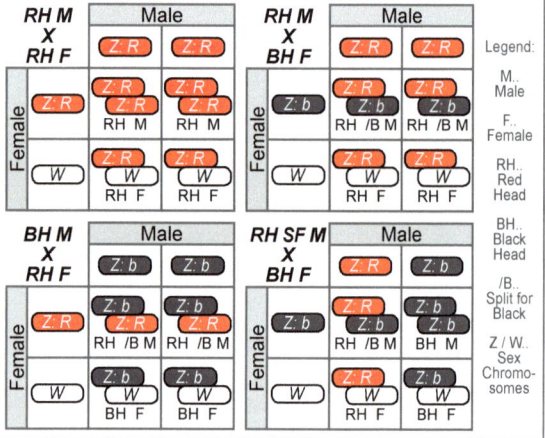

Examples of sex-linked inheritance for head colour

Sex-Linked Mutations are carried on the Z Chromosome

Females always inherit their one Z chromosome from their father

Females pass their Z chromosome to all of their sons

Females pass their Z chromosome to none of their daughters

Females determine the sex of all offspring with either a Z or a W

Impact of Single & Double Factors

In the case of Gouldian finches, head colour is sex-linked. If a colour mutation is recessive and located on the Z chromosome, a female finch (ZW) only requires one copy of the gene on her single Z chromosome to express the trait, whereas a male finch (ZZ) would need two copies to display the same mutation. This difference in inheritance patterns can result in distinct colour variations between male and female finches.

It also explains how a male can be said to be "single factor" for a recessive trait, while a female is only ever single factor for a sex-linked trait and will always express it in the phenotype or appearance.

Understanding sex-linked inheritance is crucial for breeders aiming to predict the appearance of offspring and to develop specific colour traits within their breeding programs. This knowledge allows for more precise genetic planning and the ability to achieve and maintain the desired characteristics across generations.

Co-dominant Sex-Linked Inheritance

Sex-linked inheritance can account for various traits beyond the Gouldian finch's head colours of red and black. This phenomenon occurs when two distinct alleles of a gene are situated on the sex chromosomes, and both alleles are equally expressed in the bird's phenotype. Consequently, neither allele is dominant or recessive to one another; rather, they both contribute to the observable characteristics.

An example is the Yellow Back (European YB) Gouldian, where this inheritance is a dominant sex-linked trait alongside the gene for a potential red head colour. Understanding co-dominant sex-linked inheritance is important for breeders who aim to predict and develop these mixed traits within their breeding programs.

Autosomal Inheritance

Autosomal inheritance refers to the transmission of traits

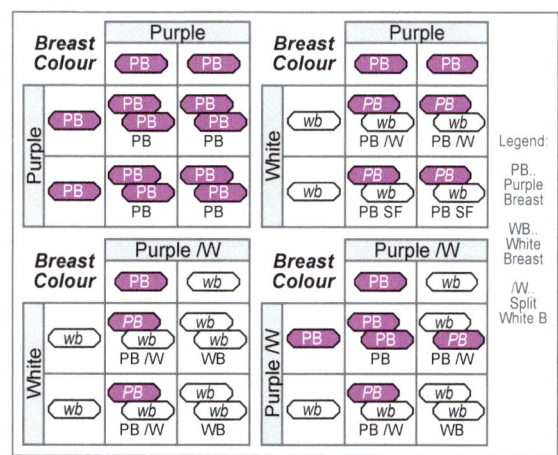

Examples of autosomal inheritance

determined by genes located on the autosomes, which are the non-sex chromosomes. In Gouldian finches, as in many other species, these autosomes come in pairs, with one chromosome inherited from each parent. This type of inheritance contrasts with sex-linked inheritance because autosomal genes are present in both males and females in equal numbers.

Just like sex-linked inheritance, traits controlled by autosomal genes can be either dominant or recessive. A dominant trait requires only one copy of the gene to be expressed in the phenotype, meaning that if an individual inherits the dominant allele from one parent, the dominant trait will be visible. On the other hand, a recessive trait requires two copies of the gene, one from each parent, to be expressed.

If an individual has only one copy of a recessive allele, they are considered a carrier or "split" and will not show the trait in their appearance, but they can pass it on to their offspring.

Understanding autosomal inheritance is essential for breeders because it allows them to predict the likelihood of certain traits appearing in the next generation. By analysing the genetic makeup of breeding pairs and using tools such as Punnett squares, breeders can estimate the probability of different genetic outcomes and make informed decisions to achieve specific breeding goals.

Breast Colour as an Example

The breast colour of Gouldian finches is a trait that appears in the phenotype of many different mutations and it is an excellent example of autosomal inheritance. Purple is dominant to lilac and lilac is dominant to white. Or stated in the reverse, white is recessive to lilac and lilac is recessive to purple.

As noted earlier, exceptions include the Australian Yellow and Lutino where the white breast colour is inherited as a trait of the mutation. Nevertheless, these still carry genes

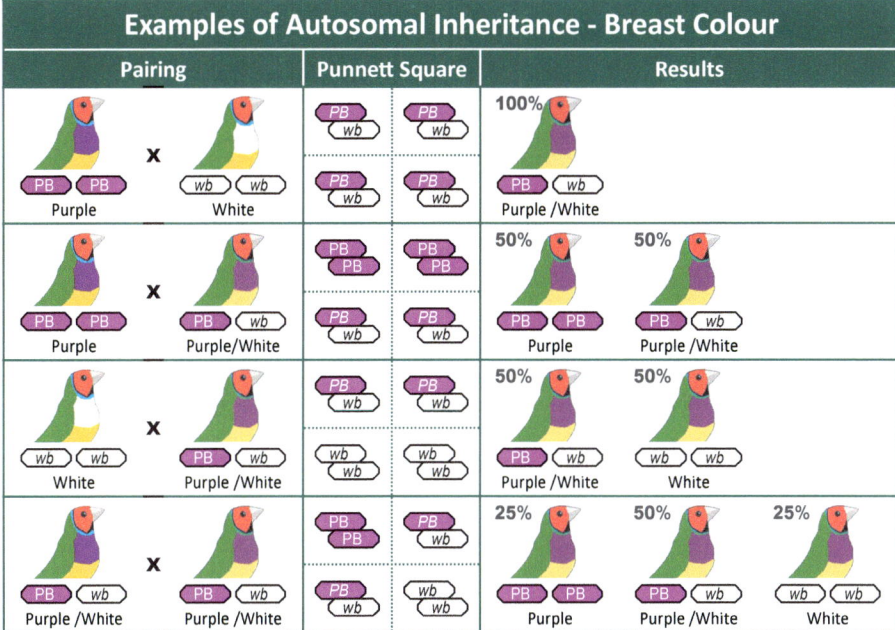

For a full summary see "Breast Colour Inheritance" on page 156

for breast colour even if the appearance is masked by the mutation (epistatic).

The breast colour inheritance results can therefore be overlaid on most of the Gouldian mutations.

Genetics of Mutations

Phenotype Abbreviations

The lists of abbreviations for the colour appearance of head, breast and back for Gouldian finches (phenotype) appear earlier in this chapter.

Genotype Abbreviations

The abbreviations for genotypes (genetics) also appear in the tables earlier in this chapter. As noted previously, the convention is that upper case indicates dominant inheritance and lower case is recessive.

While there are a large number of possible genetic combinations in Gouldian finches, the most common ones are summarised in the table titled "Quick Reference Table - Genetics of Gouldian Finches" on page 139.

Tip of Beak Colours

Another variation in the phenotypes of Gouldian finches is the tip of beak colour. This can sometimes to be used to indicate differences in the genotype of two birds, even though they have an identical appearance in every other way.

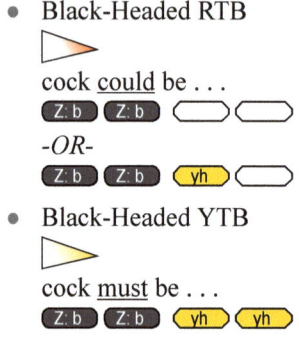

This information is useful when selecting pairs of Gouldian finches for breeding.

For example, if considering two black-headed birds with Tip of Beak of different colours it is possible to identify which one is carrying the recessive gene for Yellow Headed.

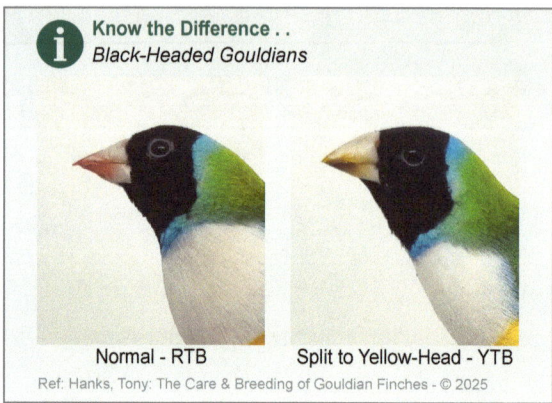

Genetic Prediction Software

With approximately 2,000 combinations possible, many breeders find it useful to use an online genetic predictor for the progeny of Gouldian finch pairings. The following are examples:

- https://www.gouldamadinecalculator.nl/en/gouldianfinch_calculator.html This site is recommended for its' less obtrusive advertising.
- http://www.gouldianfinches.eu/en/genetics-forecast/

Genetic Considerations

Quick Reference Table - Genetics of Gouldian Finches

Description	Notes	Cocks (Males) Phenotype H B K C N L T	Cocks (Males) Genotype Sex Linked	Cocks (Males) Genotype Autosomal	Hens (Females) Phenotype H B K C N L T	Hens (Females) Genotype Sex Linked	Hens (Females) Genotype Autosomal
Red Headed Normal	• Wild-type • Red TB	R P G B B Y B	Z:R / Z:R		R M G B / B L B / G Y G	Z:R / W	
Black Headed Normal	• Wild-type • Red TB	B P G B B Y B	Z:b / Z:b		B M G B / B L B / G Y G	Z:b / W	
Yellow Headed Normal	• Wild-type • Yellow TB • (Split to Red)	Y P G B B Y B	Z:R / Z:R	yh / yh	Y M G B / B L B / G Y G	Z:R / W	yh / yh
Red Headed Split to Black	• Wild-type • Red TB	R P G B B Y B	Z:R / Z:b		*A female having Red on her only Z sex chromosome cannot also be split to Black*		
Red Headed Split to Yellow	• Wild-type • Red TB	R P G B B Y B	Z:R / Z:R	yh	R M G B / B L B / G Y G	Z:R / W	yh
Red Headed Split to Black & Yellow	• Wild-type • Red TB	R P G B B Y B	Z:R / Z:b	yh	*A female having Red on her only Z sex chromosome cannot also be split to Black*		
Black Headed Split to SF Yellow	• Wild-type • Red TB	B P G B B Y B	Z:b / Z:b	yh	B M G B / B L B / G Y G	Z:b / W	yh
Black Headed Split to DF Yellow	• Wild-type • Yellow TB	B P G B B Y B	Z:b / Z:b	yh / yh	B M G B / B L B / G Y G	Z:b / W	yh / yh
Yellow Headed Split to Red & Black	• Wild-type • Yellow TB	Y P G B B Y B	Z:R / Z:b	yh / yh	*A female having Red on her only Z sex chromosome cannot also be split to Black*		

Legend - Phenotype

H	Head	T	Tail	D	Dilute	LG	Light Grey	M Mauve	VLY V Light Yellow
B	Breast	TB	Tip of Beak	DB	Dark Blue	LM	Light Mauve	OG Olive Green	W White
K	Back			G	Green	LO	Light Olive	P Purple	Y Yellow
C	Chin			B	Black	LR	Light Red	R Red	YG Yellow-Green
N	Nape			B	Blue	LS	Light Salmon	S Salmon	
L	Belly			BG	Blue-Green	LB	Light Blue		
				GB	Grey / Blue	LY	Light Yellow	VLG V Light Grey	
				G	Grey				

(Gene abbreviations are as shown earlier in this chapter)

Gouldian Finches - Care, Breeding & Genetics

Quick Reference Table - Genetics of Gouldian Finches

Description	Notes	Cocks (Males) Phenotype H B K C N L T	Cocks (Males) Genotype Sex Linked	Cocks (Males) Genotype Autosomal	Hens (Females) Phenotype H B K C N L T	Hens (Females) Genotype Sex Linked	Hens (Females) Genotype Autosomal
Red Headed White Breasted	• Mutation • Red TB	R W G B B Y B	Z:R / Z:R	wb / wb	R W G B / B L B / G Y G	Z:R / W	wb / wb
Black Headed White Breasted	• Mutation • Red TB	B W G B B Y B	Z:b / Z:b	wb / wb	B W G B / B L B / G Y G	Z:b / W	wb / wb
Yellow Headed White Breasted	• Mutation • Yellow TB	Y W G B B Y B	Z:R / Z:R	yh / yh / wb / wb	Y W G B / B L B / G Y G	Z:R / W	yh / yh / wb / wb
Red Headed Yellow Back	• Mutation • Red TB • (Euro YB) • (Pastel)	R P Y W W Y W	Z:R / Z:R / Z:YB / Z:YB		R M Y W W L Y W	Z:R / W / Z:YB / W	
Black Headed Yellow Back	• Mutation • Red TB	L G P Y W W Y W	Z:b / Z:b / Z:YB / Z:YB		L G M Y W W L Y W	Z:b / W / Z:YB / W	
Yellow Headed Yellow Back	• Mutation • Yellow TB	Y P Y W W Y W	Z:R / Z:R / Z:YB / Z:YB	yh / yh	Y M Y W W L Y W	Z:R / W / Z:YB / W	yh / yh
Red Headed Yellow Back SF Dilute Cock	• Mutation • Red TB • (Dilute) • (SF Pastel)	R P D G B Y B	Z:R / Z:R / Z:YB / —		*A single-factor sex-linked mutation appears as the standard Yellow Back in females*		
Black Headed Yellow Back SF Dilute Cock	• Mutation • Red TB	G P D G B Y B	Z:b / Z:b / Z:YB / —		*A single-factor sex-linked mutation appears as the standard Yellow Back in females*		
Yellow Head Yellow Back SF Dilute Cock	• Mutation • Yellow TB	Y P D G B Y B	Z:R / Z:R / Z:YB / Z:YB	yh / yh	*A single-factor sex-linked mutation appears as the standard Yellow Back in females*		
Red Headed Yellow Back White Breasted	• Mutation • Red TB • (Euro YB) • (Pastel)	R W Y W W Y W	Z:R / Z:R / Z:YB / Z:YB	wb / wb	R W Y W W L Y W	Z:R / W / Z:YB / W	wb / wb
Black Headed Yellow Back White Breasted	• Mutation • Red TB	L G W Y W W Y W	Z:b / Z:b / Z:YB / Z:YB	wb / wb	L G W Y W W L Y W	Z:b / W / Z:YB / W	wb / wb
Yellow Head Yellow Back White Breasted	• Mutation • Yellow TB	Y W Y W W Y W	Z:R / Z:R / Z:YB / Z:YB	yh / yh / wb / wb	Y W Y W W L Y W	Z:R / W / Z:YB / W	yh / yh / wb / wb

Genetic Considerations

Quick Reference Table - Genetics of Gouldian Finches

Description	Notes	Cocks (Males) Phenotype (H B K C N L T)	Cocks Genotype (Sex Linked / Autosomal)	Hens (Females) Phenotype (H B K C N L T)	Hens Genotype (Sex Linked / Autosomal)
Red Headed SF Y. Back White Breast Cock	• Mutation • Red TB • (Dilute) • (SF Pastel)	R W Y L/G L/G Y L/G	Z:R / Z:R / Z:YB ; wb / wb	A single-factor sex-linked mutation appears as the standard Yellow Back in females	
Black Headed SF Y. Back White Breast Cock	• Mutation • Red TB	L/G W Y L/G L/G Y L/G	Z:b / Z:b / Z:YB ; wb / wb	A single-factor sex-linked mutation appears as the standard Yellow Back in females	
Yellow Head SF Y. Back White Breast Cock	• Mutation • Yellow TB	Y W Y L/G L/G Y L/G	Z:R / Z:R / Z:YB ; yh / yh / wb / wb	A single-factor sex-linked mutation appears as the standard Yellow Back in females	
Red Headed Australian Yellow	• Mutation • Red TB • (AY)	R W Y W W Y W	Z:R / Z:R ; ay / ay	R W Y W W L/Y W	Z:R / W ; ay / ay
Black Headed Australian Yellow	• Mutation • Red TB	V L/G W Y W W Y W	Z:b / Z:b ; ay / ay	V L/G W Y W W L/Y W	Z:b / W ; ay / ay
Yellow Headed Australian Yellow	• Mutation • Yellow TB	Y W Y W W Y W	Z:R / Z:R ; yh / yh / ay / ay	Y W Y W W L/Y W	Z:R / W ; yh / yh / ay / ay
Red Headed Blue Backed	• Mutation • White BT	S P D/B B B W B	Z:R / Z:R ; bk / bk	S M D/B D/B W B	Z:R / W ; bk / bk
Black Headed Blue Backed	• Mutation • White BT	B P D/B B B W B	Z:b / Z:b ; bk / bk	B M D/B D/B W B	Z:b / W ; bk / bk
Yellow Headed Blue Backed	• Mutation • White BT • (Y.Head is not seen)	S P D/B B B W B	Z:R / Z:R ; yh / yh / bk / bk	S M D/B D/B W B	Z:R / W ; yh / yh / bk / bk
Red Headed Blue Backed White Breasted	• Mutation • White BT	L/S W D/B D/B W B	Z:R / Z:R ; bk / wb / wb	L/S W D/B D/B W B	Z:R / W ; bk / wb / wb
Black Headed Blue Backed White Breasted	• Mutation • White BT	B W D/B D/B W B	Z:b / Z:b ; bk / wb / wb	B W D/B D/B W B	Z:b / W ; bk / wb / wb
Yellow Headed Blue Backed White Breast	• Mutation • White BT • (Phenotype same as RH)	L/S W D/B D/B W B	Z:R / Z:R ; yh / yh / bk / bk / wb / wb	L/S W D/B D/B W B	Z:R / W ; yh / yh / bk / bk / wb / wb

Gouldian Finches - Care, Breeding & Genetics

Quick Reference Table - Genetics of Gouldian Finches

Description	Notes	Cocks (Males) Phenotype (H B K C N L T)	Cocks (Males) Genotype (Sex Linked / Autosomal)	Hens (Females) Phenotype (H B K C N L T)	Hens (Females) Genotype (Sex Linked / Autosomal)
Red Headed Pastel Blue SF Yellow Back Cock	• Secondary Mutation • White BT	L/S, P, L/B, G, B, W, B	Z: R / Z: R / Z: YB ; bk / bk	*A single-factor sex-linked YB mutation appears as the Silver in females*	
Black Headed Pastel Blue SF Yellow Back Cock	• Secondary Mutation • White BT	G, P, L/B, G, B, W, B	Z: b / Z: b / Z: YB ; bk / bk	*A single-factor sex-linked YB mutation appears as the Silver in females*	
Yellow Head Pastel Blue SF Yellow Back Cock	• Secondary Mutation • (Phenotype same as RH)	L/S, P, L/B, G, B, W, B	Z: R / Z: R / Z: YB ; yh / yh / bk / bk	*A single-factor sex-linked YB mutation appears as the Silver in females*	
Red Headed Silver	• Secondary Mutation • White BT	L/S, V/M, L/G, V/W, W, V/W/Y, L/W	Z: R / Z: R / Z: YB / Z: YB ; bk / bk	L/S, L/M, V/G, V/L/G, V/L/G, V/L/G/Y, L/W	Z: R / W ; Z: YB / W ; bk / bk
Black Headed Silver	• Secondary Mutation • White BT	V/L/G, V/M/G, L/W, V/W, W, V/W/Y, L/W	Z: b / Z: b ; Z: YB / Z: YB ; bk / bk	V/L/G, V/L/M, V/G, V/L/G, V/L/G, V/L/G/Y, L/W	Z: b / W ; Z: YB / W ; bk / bk
Yellow Headed Silver	• Secondary Mutation • (Phenotype same as RH)	L/S, V/M, L/G, V/W, W, V/W/Y, L/W	Z: R / Z: R / Z: YB / Z: YB ; yh / yh ; bk / bk	L/S, L/M, V/G, V/L/G, V/L/G, V/L/G/Y, L/W	Z: R / W ; yh / yh ; Z: YB / W ; bk / bk
Red Headed Silver White Breasted	• Secondary Mutation • White BT	L/S, V/W/G, L/W, V/W, W, V/W/Y, L/W	Z: R / Z: R / Z: YB ; bk / bk / wb / wb	L/S, V/W, V/L/G, V/L/G, V/L/G, V/L/G/Y, L/W	Z: R / W ; Z: YB / W ; bk / bk ; wb / wb
Black Headed Silver White Breasted	• Secondary Mutation • White BT	V/L/G, V/W/G, L/W, V/W, W, V/W/Y, L/W	Z: b / Z: b ; Z: YB / Z: YB ; bk / bk ; wb / wb	V/L/G, V/W, V/L/G, V/L/G, V/L/G, V/L/G/Y, L/W	Z: b / W ; Z: YB / W ; bk / bk ; wb / wb
Yellow Head Silver White Breasted	• Secondary Mutation • (Phenotype same as RH)	L/S, V/W/G, L/W, V/W, W, V/W/Y, L/W	yh / yh ; Z: R / Z: R ; Z: YB / Z: YB ; bk / bk ; wb / wb	L/S, V/W, V/L/G, V/L/G, V/L/G, V/L/G/Y, L/W	yh / yh ; Z: R / W ; bk / bk ; Z: YB / W ; wb / wb
Red Headed Silver White Breasted (SF Yellow Back)	• Secondary Mutation • White BT	L/S, V/W/G, L/W, V/W, W, V/W/Y, L/W	Z: R / Z: R / Z: YB ; bk / bk / wb / wb	*A single-factor sex-linked YB mutation appears as the standard Silver in females*	
Black Headed Silver White Breasted (SF Yellow Back)	• Secondary Mutation • White BT	V/L/G, V/W/G, L/W, V/W, W, V/W/Y, L/W	Z: b / Z: b ; Z: YB ; bk / bk ; wb / wb	*A single-factor sex-linked YB mutation appears as the standard Silver in females*	

Genetic Considerations

Quick Reference Table - Genetics of Gouldian Finches

Description	Notes	Cocks (Males) Phenotype H B K C N L T	Cocks (Males) Genotype Sex Linked / Autosomal	Hens (Females) Phenotype H B K C N L T	Hens (Females) Genotype Sex Linked / Autosomal
Yellow Head Silver White Breasted (SF Yellow Back)	• Secondary Mutation • (Phenotype same as RH)	LS / W / V LG / WW / V LY / W	Z: R / Z: R / Z: YB ; yh / yh / bk / bk / wb / wb	*A single-factor sex-linked YB mutation appears as the standard Silver in females*	
Red Headed Australian Recessive Dilute	• Mutation • Red TB	R P Y YG G G Y W	Z: R / Z: R ; ad / ad	R M Y YG G G Y W	Z: R / W ; ad / ad
Black Headed Australian Recessive Dilute	• Mutation • Red TB	LG P Y YG G G Y W	Z: b / Z: b ; ad / ad	LG M Y YG G G Y W	Z: b / W ; ad / ad
Yellow Head Australian Recessive Dilute	• Mutation • Yellow TB	Y P Y YG G G Y W	Z: R / Z: R ; yh / yh / ad / ad	Y M Y YG G G Y W	Z: R / W ; yh / yh / ad / ad
Red Headed Australian Recess Dilute White Breast	• Mutation • Red TB	R W Y YG G G Y W	Z: R / Z: R ; ad / ad / wb / wb	R W Y YG G G Y W	Z: R / W ; ad / ad / wb / wb
Black Headed Australian Recess Dilute White Breast	• Mutation • Red TB	LG W Y YG G G Y W	Z: b / Z: b ; ad / ad / wb / wb	LG W Y YG G G Y W	Z: b / W ; ad / ad / wb / wb
Yellow Head Australian Recess Dilute White Breast	• Mutation • Yellow TB	Y W Y YG G G Y W	Z: R / Z: R ; yh / yh / ad / ad / wb / wb	Y W Y YG G G Y W	Z: R / W ; yh / yh / ad / ad / wb / wb
Red Headed Australian Recessive Dilute Blue	• Mutation • Red TB	R P G GB L LBY L G Y G	Z: R / Z: R ; ad / ad / bk / bk	R M G GB L LBY G Y G	Z: R / W ; ad / ad / bk / bk
Black Headed Australian Recessive Dilute Blue	• Mutation • Red TB	G P G GB L LBY G Y G	Z: b / Z: b ; ad / ad / bk / bk	G M G GB L LBY G Y G	Z: b / W ; ad / ad / bk / bk
Yellow Head Australian Recessive Dilute Blue	• Mutation • Yellow TB	Y P G GB L LBY G Y G	Z: R / Z: R ; yh / yh / ad / ad / bk / bk	Y M G GB L LBY G Y G	Z: R / W ; yh / yh / ad / ad / bk / bk
Red Headed Cinnamon	• Mutation • Red TB	R P O OG G B Y G	Z: R / Z: R / Z: cn / Z: cn	R M O OG G Y G	Z: R / W / Z: cn / W

Gouldian Finches - Care, Breeding & Genetics

Quick Reference Table - Genetics of Gouldian Finches

Description	Notes	Cocks (Males) Phenotype H B K C N L T	Cocks (Males) Genotype Sex Linked	Cocks (Males) Genotype Autosomal	Hens (Females) Phenotype H B K C N L T	Hens (Females) Genotype Sex Linked	Hens (Females) Genotype Autosomal
Black Headed Cinnamon	• Mutation • Red TB	G P O/G O G B Y G	Z: b / Z: b / Z: cn / Z: cn		G M O/G O G Y G	Z: b / W / Z: cn / W	
Yellow Headed Cinnamon	• Mutation • Yellow TB	Y P O/G O G B Y G	Z: R / Z: R / Z: cn / Z: cn	yh / yh	Y M O/G O G Y G	Z: R / W / Z: cn / W	yh / yh
Red Headed Seagreen	• Mutation • Red TB	L/R P B/G B B L/Y B	Z: R / Z: R / Z: sg / Z: sg		L/R M B/G B B L/Y B	Z: R / W / Z: sg / W	
Black Headed Seagreen	• Mutation • Red TB	B P B/G B B L/Y B	Z: b / Z: b / Z: sg / Z: sg		B M B/G B B L/Y B	Z: b / W / Z: sg / W	
Yellow Headed Seagreen	• Mutation • Yellow TB	L/Y P B/G B B L/Y B	Z: R / Z: R / Z: sg / Z: sg	yh / yh	L/Y M B/G B B L/Y B	Z: R / W / Z: sg / W	yh / yh
Red Headed Fallow	• Mutation • Red TB • (AY)	L/R O/M G L/B L O/Y G	Z: R / Z: R	fa / fa	L/R L/M O G G/B L/Y O/G	Z: R / W	fa / fa
Black Headed Fallow	• Mutation • Red TB	G O/M G L/B L O/Y G	Z: b / Z: b	fa / fa	G L/M O G G/B L/Y O/G	Z: b / W	fa / fa
Yellow Headed Fallow	• Mutation • Yellow TB	L/Y O/M G L/B L O/Y G	Z: R / Z: R	yh / yh / fa / fa	L/Y L/M O G G/B L/Y O/G	Z: R / W	yh / yh / fa / fa
Red Headed Lutino	• Mutation • Red TB	W W Y W W Y W	Z: R / Z: R / Z: ino / Z: ino		W W Y W W Y W	Z: R / W / Z: ino / W	
Black Headed Lutino	• Mutation • (Phenotype same as RH)	W W Y W W Y W	Z: b / Z: b / Z: ino / Z: ino		W W Y W W Y W	Z: b / W / Z: ino / W	

Legend - Phenotype

H Head	T Tail	D Dilute	LG Light Grey	M Mauve	VLY V Light Yellow
B Breast	TB Tip of Beak	DB Dark Blue	LM Light Mauve	OG Olive Green	W White
K Back		G Green	LO Light Olive	P Purple	Y Yellow
C Chin	B Black	G Grey	LR Light Red	R Red	YG Yellow-Green
N Nape	B Blue	GB Grey / Blue	LS Light Salmon	S Salmon	
L Belly	BG Blue-Green	LB Light Blue	LY Light Yellow	VLG V Light Grey	

(Gene abbreviations are as shown earlier in this chapter)

Genetic Considerations

Quick Reference Table - Genetics of Gouldian Finches

Description		Notes	Cocks (Males) Phenotype (H B K C N L T)	Cocks (Males) Genotype Sex Linked	Cocks (Males) Genotype Autosomal	Hens (Females) Phenotype (H B K C N L T)	Hens (Females) Genotype Sex Linked	Hens (Females) Genotype Autosomal
Yellow Headed Lutino		• Mutation • (Phenotype same as RH)	W W Y W W Y W	Z: R / Z: R / Z: ino / Z: ino	yh / yh	W W Y W W Y W	Z: R / W / Z: ino / W	yh / yh
Red Headed Ivory		• Secondary Mutation • White BT	R W B/G B/G L/Y L G	Z: R / Z: R / Z: cn / Z: cn	bk / bk	R W B/G B/G L/Y L G	Z: R / W / Z: cn / W	bk / bk
Black Headed Ivory		• Secondary Mutation • White BT	G W B/G B/G L/Y L G	Z: b / Z: b / Z: cn / Z: cn	bk / bk	G W B/G B/G L/Y L G	Z: b / W / Z: cn / W	bk / bk
Yellow Headed Ivory		• Secondary Mutation • White BT	Y W B/G B/G L/Y L G	Z: R / Z: R / Z: cn / Z: cn	yh / yh / bk / bk	Y W B/G B/G L/Y L G	Z: R / W / Z: cn / W	yh / yh / bk / bk
Red Headed Satine		• Secondary Mutation • Red TB	L/O W L/O L/O G G L/Y W	Z: R / Z: R / Z: cn / Z: cn / Z: ino / Z: ino		L/O W L/O L/O G G L/Y W	Z: R / W / Z: cn / W / Z: ino / W	
Black Headed Satine		• Secondary Mutation • (Phenotype same as RH)	L/O W L/O L/O G G L/Y W	Z: b / Z: b / Z: cn / Z: cn / Z: ino / Z: ino		L/O W L/O L/O G G L/Y W	Z: b / W / Z: cn / W / Z: ino / W	
Yellow Headed Satine		• Secondary Mutation • (Phenotype same as RH)	L/O W L/O L/O G G L/Y W	Z: R / Z: R / Z: cn / Z: cn / Z: ino / Z: ino	yh / yh	L/O W L/O L/O G G L/Y W	Z: R / W / Z: cn / W / Z: ino / W	yh / yh
Red Headed Australian Variegated Blue (AVB)		• Secondary Mutation • White BT	L/S W G/B W W W W	Z: R / Z: R	ay / ay / bk / bk	L/S W G/B W W W W	Z: R / W	ay / ay / bk / bk
Black Headed Australian Variegated Blue (AVB)		• Secondary Mutation • White BT	G W G/B W W W W	Z: R / Z: R	ay / ay / bk / bk	G W G/B W W W W	Z: b / W	ay / ay / bk / bk
Yellow Head Australian Variegated Blue (AVB)		• Secondary Mutation • (Phenotype same as RH)	L/S W G/B W W W W	Z: R / Z: R	yh / yh / ay / ay / bk / bk	L/S W G/B W W W W	Z: R / W	yh / yh / ay / ay / bk / bk

© 2025 - Hanks, Tony

Progeny Predictions

Understanding these patterns of inheritance can clarify certain breeding outcomes that might seem unexpected.

Punnett Squares for 1 Trait

One of the easiest ways to visualise progeny predictions is by using Punnett Squares to show how alleles segregate and combine during reproduction. Once you understand the principles you can quickly make any prediction using this method.

For instance, breeding a Red-Headed Cock with a Blacked-Head Hen will produce the surprising result where the offspring will all having Red-Heads. Although they will have split chromosomes in their genotypes, their phenotypes will uniformly display Red-Heads.

		COCK	
		Red Head Normal	
		Z_R	Z_R
HEN / Black Head Normal	Z_b	$Z_R Z_b$	$Z_R Z_b$
HEN / Black Head Normal	W	$Z_R W$	$Z_R W$

Progeny Expected:
- 50% RH Cocks split BH
- 50% RH Hens

How to Use Punnett Squares

Using a Punnett Square to calculate the predicted outcomes is not difficult. The example on the following page demonstrates the steps involved. (The form for Punnett Squares is contained in the "Forms Library" of this book - see page 203).

Genetic Considerations

Four Steps to Use a Punnett Square

①

	COCK *Red Head split Black*	
	Z	Z
HEN *Black Head normal* — Z	Z Z	Z Z
W	Z W	Z W

Progeny Expected:

1: Record the appearances (phenotypes)

②

	COCK *Red Head split Black*	
	Z R	Z b
HEN *Black Head normal* — Z b	Z Z	Z Z
W	Z W	Z W

Progeny Expected:

2: Record the genes (genotypes)

③

	COCK *Red Head split Black*	
	Z R	Z b
HEN *Black Head normal* — Z b	Z R Z b	Z b Z b
W	Z R W	Z b W

Progeny Expected:

3: Note the intersecting pairs of genes

④

	COCK *Red Head split Black*	
	Z R	Z b
HEN *Black Head normal* — Z b	Z R Z b	Z b Z b
W	Z R W	Z b W

Progeny Expected:
 25% RH cocks / black
 25% BH cocks
 25% RH hens
 25% BH hens

4: Interpret the genes into their phenotypes

Example of using a Punnett Square for progeny prediction

Examples of Calculated Outcomes

The following graphics show examples of how to visualise expected outcomes:

COCK		
Red Head spit Black		
	Z R	Z b
HEN Red Head Normal — Z R	Z R Z R	Z R Z b
HEN Red Head Normal — W	Z R W	Z b W

Progeny Expected:
- 25% RH Cocks
- 25% RH Cocks split Black
- 25% RH Hens
- 25% BH Hens

COCK		
Red Head split Black		
	Z R	Z b
HEN Black Head Normal — Z b	Z R Z b	Z b Z b
HEN Black Head Normal — W	Z R W	Z b W

Progeny Expected:
- 25% RH Cocks split Black
- 25% BH Cocks
- 25% RH Hens
- 25% BH Hens

Double Punnett Squares

Often a progeny prediction will involve more than one trait. This is called a dihybrid cross, where there are two traits determined by a separate pair of alleles on different chromosomes. For example, a white breasted yellow back Gouldian. This requires a larger Punnett Square, but it is still not difficult.

As an example, consider a pairing of an Australian Yellow cock with a White Breasted Yellow Back hen. It is not normally recommended to mix these mutations because it will be difficult to feel confident about the genetics of the offspring, but it is a useful example.

			COCK			
			Australian Yellow			
			Z Z ay ay (PB PB)			
			Z ay PB	Z ay PB	Z ay PB	Z ay PB
HEN White Breast Yellow Back	Z YB W	wb Z YB	Z ay PB wb Z YB	Z ay PB wb Z YB	Z ay PB wb Z YB	Z ay PB wb Z YB
		wb W	Z ay PB wb W	Z ay PB wb W	Z ay PB wb W	Z ay PB wb W
	wb wb	wb Z YB	Z ay PB wb Z YB	Z ay PB wb Z YB	Z ay PB wb Z YB	Z ay PB wb Z YB
		wb W	Z ay PB wb W	Z ay PB wb W	Z ay PB wb W	Z ay PB wb W

Progeny Expected:
- 50% Purple Breasted Yellow Back
- 50% Australian Yellow

While the Australian Yellow will display a white breast as part of that mutation, it will still have breast colour information in the genotype - in this example it is assumed that the masked genes for breast colour are the most common, purple:

While both parents had white breasts, 50% of the progeny in this example will be purple breasted. This is because the gene for purple breasted was being masked in the cock due to the pigment changes caused by the Australian Yellow mutation.

Reference Tables for Progeny Predictions

As an alternative to Genetic Prediction Software or Punnett Squares is to simply look up the expected progeny in the following table on page 150. First start with the cock and then select the description of the hen in the pairing.

Gouldian Finches - Care, Breeding & Genetics

Quick Reference Table - Gouldian Finch Progeny

Cock: Red Headed [PB] [GK] — Normal "Wild-Types"

Hen	Progeny Expected	
Red Headed [PB] [GK]	• 50% CK Red Headed	[PB] [GK]
	• 50% HN Red Headed	[PB] [GK]
Black Headed [PB] [GK]	• 50% CK Red Headed	[PB] [GK]
	• 50% HN Black Headed	[PB] [GK]
Yellow Headed [PB] [GK]	• 50% CK Red Headed (split Yellow SF)	[PB] [GK] (/ •)
	• 50% HN Red Headed (split Yellow SF)	[PB] [GK] (/ •)
Red Headed (split Yellow SF) [PB] [GK] (/ •)	• 25% CK Red Headed	[PB] [GK]
	• 25% CK Red Headed (split Yellow SF)	[PB] [GK] (/ •)
	• 25% HN Red Headed	[PB] [GK]
	• 25% HN Red Headed (split Yellow SF)	[PB] [GK] (/ •)
Black Headed (split Yellow SF) [PB] [GK] (/ •)	• 25% CK Red Headed (split Black)	[PB] [GK] (/ •)
	• 25% CK Red Headed (split Black / Yellow SF)	[PB] [GK] (/ • •)
	• 25% HN Red Headed	[PB] [GK]
	• 25% HN Red Headed (split Yellow SF)	[PB] [GK] (/ •)
Black Headed (split Yellow DF) [PB] [GK] (/ • •)	• 50% CK Red Headed (split Black / Yellow SF)	[PB] [GK] (/ • •)
	• 50% HN Red Headed (split Yellow SF)	[PB] [GK] (/ •)

Cock: Yellow Headed [PB] [GK] — Normal "Wild-Types"

Hen	Progeny Expected	
Red Headed [PB] [GK]	• 50% CK Red Headed (split Black)	[PB] [GK] (/ •)
	• 50% HN Black Headed	[PB] [GK]
Black Headed [PB] [GK]	• 50% CK Black Headed	[PB] [GK]
	• 50% HN Black Headed	[PB] [GK]
Yellow Headed [PB] [GK]	• 50% CK Red Headed (split Black, split Yellow SF)	[PB] [GK] (/ • •)
	• 50% HN Black Headed (split Yellow SF)	[PB] [GK] (/ •)
Red Headed (split Yellow SF) [PB] [GK] (/ •)	• 25% CK Red Headed (split Black)	[PB] [GK] (/ •)
	• 25% CK Red Headed (split Black / Yellow SF)	[PB] [GK] (/ • •)
	• 25% HN Black Headed	[PB] [GK]
	• 25% HN Black Headed (split Yellow SF)	[PB] [GK] (/ •)
Black Headed (split Yellow SF) [PB] [GK] (/ •)	• 25% CK Black Headed	[PB] [GK]
	• 25% CK Black Headed (split Yellow SF)	[PB] [GK] (/ •)
	• 25% HN Black Headed	[PB] [GK]
	• 25% HN Black Headed (split Yellow SF)	[PB] [GK] (/ •)
Black Headed (split Yellow DF) [PB] [GK] (/ • •)	• 50% CK Black Headed (split Yellow SF)	[PB] [GK] (/ •)
	• 50% HN Black Headed (split Yellow SF)	[PB] [GK] (/ •)

Genetic Considerations

Quick Reference Table - Gouldian Finch Progeny

Normal "Wild-Types"

Cock		Hen	Progeny Expected
Yellow Headed	PB GK		
x			
		Hen	**Progeny Expected**
Red Headed	PB GK		• 50% CK Red Headed (split Yellow SF) PB GK (/ •)
			• 50% HN Red Headed (split Yellow SF) PB GK (/ •)
Black Headed	PB GK		• 50% CK Red Headed (split Black / Yellow SF) PB GK (/ • •)
			• 50% HN Red Headed (split Yellow SF) PB GK (/ •)
Yellow Headed	PB GK		• 50% CK Yellow Headed PB GK
			• 50% HN Yellow Headed PB GK
Red Headed (split Yellow SF)	PB GK (/ •)		• 25% CK Yellow Headed PB GK
			• 25% CK Red Headed (split Yellow SF) PB GK (/ •)
			• 25% HN Yellow Headed PB GK
			• 25% HN Red Headed (split Yellow SF) PB GK (/ •)
Black Headed split Yellow SF	PB GK (/ •)		• 25% CK Yellow Headed (split Black) PB GK (/ •)
			• 25% CK Red Headed (split Black / Yellow SF) PB GK (/ • •)
			• 25% HN Yellow Headed PB GK
			• 25% HN Red Headed (split Yellow SF) PB GK (/ •)
Black Headed (split Yellow DF)	PB GK (/ • •)		• 50% CK Yellow Headed (split Black) PB GK (/ •)
			• 50% HN Yellow Headed PB GK

Normal "Wild-Types"

Cock		Hen	Progeny Expected
Black Headed (split Yellow DF)	PB GK (/ • •)		
x			
		Hen	**Progeny Expected**
Red Headed	PB GK		• 50% CK Red Headed (split Black / Yellow SF) PB GK (/ • •)
			• 50% HN Black Headed (split Yellow SF) PB GK (/ •)
Black Headed	PB GK		• 50% CK Black Headed (split Yellow SF) PB GK (/ •)
			• 50% HN Black Headed (split Yellow SF) PB GK (/ •)
Yellow Headed	PB GK		• 50% CK Yellow Headed (split Black) PB GK (/ •)
			• 50% HN Black Headed (split Yellow DF) PB GK (/ • •)
Red Headed (split Yellow SF)	PB GK (/ •)		• 25% CK Red Headed (split Black / Yellow SF) PB GK (/ • •)
			• 25% CK Yellow Headed (split Black) PB GK (/ •)
			• 25% HN Black Headed (split Yellow SF) PB GK (/ •)
			• 25% HN Black Headed (split Yellow DF) PB GK (/ • •)
Black Headed split Yellow SF	PB GK (/ •)		• 25% CK Black Headed (split Yellow SF) PB GK (/ •)
			• 25% CK Black Headed (split Yellow DF) PB GK (/ • •)
			• 25% HN Black Headed (split Yellow SF) PB GK (/ •)
			• 25% HN Black Headed (split Yellow DF) PB GK (/ • •)
Black Headed (split Yellow DF)	PB GK (/ • •)		• 50% CK Black Headed (split Yellow DF) PB GK (/ • •)
			• 50% HN Black Headed (split Yellow DF) PB GK (/ • •)

Quick Reference Table - Gouldian Finch Progeny

Cock	
Red Headed (split Black) 🔴 PB GK (/ ●)	Normal "Wild-Types"

Hen	Progeny Expected
Red Headed 🔴 PB GK	• 25% CK Red Headed 🔴 PB GK
	• 25% CK Red Headed (split Black) 🔴 PB GK (/ ●)
	• 25% HN Red Headed 🔴 PB GK
	• 25% HN Black Headed ⚫ PB GK
Black Headed ⚫ PB GK	• 25% CK Black Headed ⚫ PB GK
	• 25% CK Red Headed (split Black) 🔴 PB GK (/ ●)
	• 25% HN Red Headed 🔴 PB GK
	• 25% HN Black Headed ⚫ PB GK
Yellow Headed 🟠 PB GK	• 25% CK Red Headed (split Yellow SF) 🔴 PB GK (/ ●)
	• 25% CK Red Headed (split Black / Yellow SF) 🔴 PB GK (/ ● ●)
	• 25% HN Red Headed (split Yellow SF) 🔴 PB GK (/ ●)
	• 25% HN Black Headed (split Yellow SF) ⚫ PB GK (/ ●)
Red Headed (split Yellow SF) 🔴 PB GK (/ ●)	• 12.5% CK Red Headed 🔴 PB GK
	• 12.5% CK Red Headed (split Black) 🔴 PB GK (/ ●)
	• 12.5% CK Red Headed (split Yellow SF) 🔴 PB GK (/ ●)
	• 12.5% CK Red Headed (split Black / Yellow SF) 🔴 PB GK (/ ● ●)
	• 12.5% HN Red Headed 🔴 PB GK
	• 12.5% HN Black Headed ⚫ PB GK
	• 12.5% HN Red Headed (split Yellow SF) 🔴 PB GK (/ ●)
	• 12.5% HN Black Headed (split Yellow SF) ⚫ PB GK (/ ●)
Black Headed split Yellow SF ⚫ PB GK (/ ●)	• 12.5% CK Black headed ⚫ PB GK
	• 12.5% CK Black Headed (split Yellow SF) ⚫ PB GK (/ ●)
	• 12.5% CK Red Headed (split Black) 🔴 PB GK (/ ●)
	• 12.5% CK Red Headed (split Black / Yellow SF) 🔴 PB GK (/ ● ●)
	• 12.5% HN Red Headed 🔴 PB GK
	• 12.5% HN Red Headed (split Yellow SF) 🔴 PB GK (/ ●)
	• 12.5% HN Black Headed ⚫ PB GK
	• 12.5% HN Black Headed (split Yellow SF) ⚫ PB GK (/ ●)
Black Headed (split Yellow DF) ⚫ PB GK (/ ● ●)	• 25% CK Red Headed (split Black / Yellow SF) 🔴 PB GK (/ ● ●)
	• 25% CK Black Headed (split Yellow SF) ⚫ PB GK (/ ●)
	• 25% HN Red Headed (split Yellow SF) 🔴 PB GK (/ ●)
	• 25% HN Black Headed (split Yellow SF) ⚫ PB GK (/ ●)

Genetic Considerations

Quick Reference Table - Gouldian Finch Progeny

Cock		
Red Headed (split Yellow) 🔴 PB GK (/ •)		Normal "Wild-Types"

Hen	Progeny Expected	
Red Headed 🔴 PB GK	• 25% CK Red Headed	🔴 PB GK
	• 25% CK Red Headed (split Yellow SF)	🔴 PB GK (/ •)
	• 25% HN Red Headed	🔴 PB GK
	• 25% HN Red Headed (split Yellow SF)	🔴 PB GK (/ •)
Black Headed ⚫ PB GK	• 25% CK Red Headed (split Black)	🔴 PB GK (/ •)
	• 25% CK Red Headed (split Black / Yellow SF)	🔴 PB GK (/ • •)
	• 25% HN Red Headed	🔴 PB GK
	• 25% HN Red Headed (split Yellow SF)	🔴 PB GK (/ •)
Yellow Headed 🟡 PB GK	• 25% CK Yellow Headed	🟡 PB GK
	• 25% CK Red Headed (split Yellow SF)	🔴 PB GK (/ •)
	• 25% HN Yellow Headed	🟡 PB GK
	• 25% HN Red Headed (split Yellow SF)	🔴 PB GK (/ •)
Red Headed (split Yellow SF) 🔴 PB GK (/ •)	• 12.5% CK Red Headed	🔴 PB GK
	• 25% CK Red Headed (split Yellow SF)	🔴 PB GK (/ •)
	• 12.5% CK Yellow Headed	🟡 PB GK
	• 12.5% HN Red Headed	🔴 PB GK
	• 25% HN Red Headed (split Yellow SF)	🔴 PB GK (/ •)
	• 12.5% HN Yellow Headed	🟡 PB GK
Black Headed (split Yellow SF) ⚫ PB GK (/ •)	• 12.5% CK Red Headed (split Black)	🔴 PB GK (/ •)
	• 25% CK Red Headed (split Black / Yellow SF)	🔴 PB GK (/ • •)
	• 12.5% CK Yellow Headed (split Black)	🟡 PB GK (/ •)
	• 12.5% HN Red Headed	🔴 PB GK
	• 25% HN Red Headed (split Yellow SF)	🔴 PB GK (/ •)
	• 12.5% HN Yellow Headed	🟡 PB GK
Black Headed (split Yellow DF) ⚫ PB GK (/ ••)	• 25% CK Red Headed (split Black / Yellow SF)	🔴 PB GK (/ • •)
	• 25% CK Yellow Headed (split Black)	🟡 PB GK (/ •)
	• 25% HN Red Headed (split Yellow SF)	🔴 PB GK (/ •)
	• 25% HN Yellow Headed	🟡 PB GK

Legend - Progeny

🔴 Red Head	PB Purple Breast	GK Green Back	PK Pastel Blue Back	• Split Black Head
⚫ Black Head	WB White Breast	YB Yellow Back (EU)		○ Split Yellow Hd SF
🟡 Yellow Head	AY Australian Yellow	Red Tip Beak	•• Split Yellow Hd DF
Salmon Head	SF Single Factor	SK Silver Back	Yellow Tip Beak	/ Split (various)
Lt Salmon Head	DF Double Factor	DK Dilute Back	White Tip Beak	

Gouldian Finches - Care, Breeding & Genetics **153**

Quick Reference Table - Gouldian Finch Progeny

Normal "Wild-Types"

Cock		
Black Headed (split Yellow SF)	PB GK (/ •)	

Hen	Progeny Expected
Red Headed PB GK	• 25% CK Red Headed (split Black) PB GK (/ •)
	• 25% CK Red Headed (split Black / Yellow SF) PB GK (/ • •)
	• 25% HN Red Headed PB GK
	• 25% HN Black Headed (split Yellow SF) PB GK (/ •)
Black Headed PB GK	• 25% CK Black Headed PB GK
	• 25% CK Black Headed (split Yellow SF) PB GK (/ •)
	• 25% HN Black Headed PB GK
	• 25% HN Black Headed (split Yellow SF) PB GK (/ •)
Yellow Headed PB GK	• 25% CK Yellow Headed (split Black) PB GK (/ •)
	• 25% CK Red Headed (split Black / Yellow SF) PB GK (/ • •)
	• 25% HN Black Headed (split Yellow SF) PB GK (/ •)
	• 25% HN Black Headed (split Yellow DF) PB GK (/ • •)
Red Headed (split Yellow SF) PB GK (/ •)	• 12.5% CK Red Headed (split Black) PB GK (/ •)
	• 25% CK Red Headed (split Black / Yellow SF) PB GK (/ • •)
	• 12.5% CK Yellow Headed (split Black) PB GK (/ •)
	• 12.5% HN Black Headed PB GK
	• 25% HN Black Headed (split Yellow SF) PB GK (/ •)
	• 12.5% HN Black Headed (split Yellow DF) PB GK (/ • •)
Black Headed (split Yellow SF) PB GK (/ •)	• 12.5% CK Black headed PB GK
	• 25% CK Black Headed (split Yellow SF) PB GK (/ •)
	• 12.5% CK Black Headed (split Yellow DF) PB GK (/ • •)
	• 12.5% HN Black Headed PB GK
	• 25% HN Black Headed (split Yellow SF) PB GK (/ •)
	• 12.5% HN Black Headed (split Yellow DF) PB GK (/ • •)
Black Headed (split Yellow DF) PB GK (/ • •)	• 25% CK Black Headed (split Yellow SF) PB GK (/ •)
	• 25% CK Black Headed (split Yellow DF) PB GK (/ • •)
	• 25% HN Black Headed (split Yellow SF) PB GK (/ •)
	• 25% HN Black Headed (split Yellow DF) PB GK (/ • •)

Genetic Considerations

Quick Reference Table - Gouldian Finch Progeny

Cock		
Yellow Headed (split Black)	PB GK (/ •)	**Normal "Wild-Types"**

X

Hen	Progeny Expected	
Red Headed — PB GK	• 25% CK Red Headed (split Yellow SF)	PB GK (/ •)
	• 25% CK Red Headed (split Black / Yellow SF)	PB GK (/ • •)
	• 25% HN Red Headed (split Yellow SF)	PB GK (/ •)
	• 25% HN Black Headed (split Yellow SF)	PB GK (/ •)
Black Headed — PB GK	• 25% CK Black Headed (split Yellow SF)	PB GK (/ •)
	• 25% CK Red Headed (split Black / Yellow SF)	PB GK (/ • •)
	• 25% HN Red Headed (split Yellow SF)	PB GK (/ •)
	• 25% HN Black Headed (split Yellow SF)	PB GK (/ •)
Yellow Headed — PB GK	• 25% CK Yellow Headed	PB GK
	• 25% CK Yellow Headed (split Black)	PB GK (/ •)
	• 25% HN Yellow Headed	PB GK
	• 25% HN Black Headed (split Yellow DF)	PB GK (/ • •)
Red Headed (split Yellow SF) — PB GK (/ •)	• 12.5% CK Yellow Headed	PB GK
	• 12.5% CK Red Headed (split Yellow SF)	PB GK (/ •)
	• 12.5% CK Yellow Headed (split Black)	PB GK (/ •)
	• 12.5% CK Red Headed (split Black / Yellow SF)	PB GK (/ • •)
	• 12.5% HN Yellow Headed	PB GK
	• 12.5% HN Red Headed (split Yellow SF)	PB GK (/ •)
	• 12.5% HN Black Headed (split Yellow SF)	PB GK (/ •)
	• 12.5% HN Black Headed (split Yellow DF)	PB GK (/ • •)
Black Headed (split Yellow SF) — PB GK (/ •)	• 12.5% CK Yellow Headed (split Black)	PB GK (/ •)
	• 12.5% CK Black Headed (split Yellow SF)	PB GK (/ •)
	• 12.5% CK Black Headed (split Yellow DF)	PB GK (/ • •)
	• 12.5% CK Red Headed (split Black / Yellow SF)	PB GK (/ • •)
	• 12.5% HN Yellow Headed	PB GK
	• 12.5% HN Red Headed (split Yellow SF)	PB GK (/ •)
	• 12.5% HN Black Headed (split Yellow SF)	PB GK (/ •)
	• 12.5% HN Black Headed (split Yellow DF)	PB GK (/ • •)
Black Headed (split Yellow DF) — PB GK (/ • •)	• 25% CK Yellow Headed (split Black)	PB GK (/ •)
	• 25% CK Black Headed (split Yellow DF)	PB GK (/ • •)
	• 25% HN Yellow Headed	PB GK
	• 25% HN Black Headed (split Yellow DF)	PB GK (/ • •)

Legend - Progeny

	Red Head	PB	Purple Breast	GK	Green Back	PK	Pastel Blue Back	• Split Black Head
	Black Head	WB	White Breast	YB	Yellow Back (EU)			• Split Yellow Hd SF
	Yellow Head			AY	Australian Yellow		Red Tip Beak	• • Split Yellow Hd DF
	Salmon Head	SF	Single Factor	SK	Silver Back		Yellow Tip Beak	/ Split (various)
	Lt Salmon Head	DF	Double Factor	DK	Dilute Back		White Tip Beak	

Gouldian Finches - Care, Breeding & Genetics

Quick Reference Table - Gouldian Finch Progeny

BREAST COLOUR INHERITANCE

Breast colour inheritance is autosomal, so independent of the sex chromosomes. Three different breast colours are possible in most of the mutations. The particular exception is Australian Yellow (AY) where breast colour is determined by the AY mutation itself.

Purple is dominant to Lilac and White; while Lilac is dominant to White.

Parents	Progeny
Purple × Purple	• 100% Purple PB
Purple × Purple/Lilac	• 50% Purple PB • 50% Purple/Lilac PB (/LB)
Purple × Purple/White	• 50% Purple PB • 50% Purple/Wht PB (/WB)
Purple × Lilac	• 100% Purple/Lilac PB (/LB)
Purple × Lilac/White	• 50% Purple/Lilac PB (/LB) • 50% Purple/Wht PB (/WB)
Purple × White	• 100% Purple/Wht PB (/WB)
Purple/Lilac × Purple/Lilac	• 25% Purple PB • 25% Lilac LB • 50% Purple/Lilac PB (/LB)
Purple/Lilac × Purple/White	• 25% Purple PB • 25% Purple/Lilac PB (/LB) • 25% Purple/Wht PB (/WB) • 25% Lilac/Wht LB (/WB)
Purple/Lilac × Lilac	• 50% Purple/Lilac PB (/LB) • 50% Lilac LB
Purple/Lilac × Lilac/White	• 25% Purple/Lilac PB (/LB) • 25% Purple/Wht PB (/WB) • 25% Lilac LB • 25% Lilac/Wht LB (/WB)
Purple/Lilac × White	• 50% Purple/Wht PB (/WB) • 50% Lilac/Wht LB (/WB)
Purple/White × Purple/White	• 25% Purple PB • 25% White WB • 50% Purple/Wht PB (/WB)
Purple/White × Lilac	• 25% Purple/Lilac PB (/LB) • 25% Lilac/Wht LB (/WB)
Purple/White × Lilac/White	• 25% Purple/Lilac PB (/LB) • 25% Purple/Wht PB (/WB) • 25% Lilac/Wht LB (/WB) • 25% White WB
Purple/White × White	• 50% Purple/Wht PB (/WB) • 50% White WB
Lilac × Lilac	• 100% Lilac LB
Lilac × Lilac/White	• 50% Lilac LB • 50% Lilac/Wht LB (/WB)
Lilac × White	• 100% Lilac/Wht LB (/WB)
Lilac/White × Lilac/White	• 25% Lilac LB • 25% White WB • 50% Lilac/Wht LB (/WB)
Lilac/White × White	• 50% Lilace/Wht LB (/WB) • 50% White WB
White × White	• 100% White WB
White × Purple	• 100% Purple/Wht PB (/WB)
White × Purple/Lilac	• 50% Purple/Wht PB (/WB) • 50% Lilac/Wht LB (/WB)
White × Purple/White	• 50% Purple/Wht PB (/WB) • 50% White WB
White × Lilac	• 100% Lilac/Wht LB (/WB)
White × Lilac/White	• 50% Lilace/Wht LB (/WB) • 50% White WB

Dominant → Recessive

Genetic Considerations

Quick Reference Table - Gouldian Finch Progeny

MUTATIONS - WHITE BREASTED

With over 50 different cocks and 40 different hens across all the mutations and split genes, there are approximately 2,000 possible combinations, so only the most popular combinations can be included here. Consult an online Gouldian calculators for any other pairings.

Cock		Progeny Expected	
Red Headed White Breasted WB GK			

X

Hen	Progeny Expected		
Red Headed Normal PB GK	• 50% CK Red Headed Normal (split WB)	PB GK (/ WB)	
	• 50% HN Red Headed Normal (split WB)	PB GK (/ WB)	
Red Headed (split White Breasted) PB GK (/ WB)	• 25% CK Red Headed Normal (split WB)	PB GK (/ WB)	
	• 25% CK Red Headed White Breasted	WB GK	
	• 25% HN Red Headed Normal (split WB)	PB GK (/ WB)	
	• 25% HN Red Headed White Breasted	WB GK	
Red Headed White Breasted WB GK	• 50% CK Red Headed White Breasted	WB GK	
	• 50% HN Red Headed White Breasted	WB GK	
Black Headed Normal PB GK	• 50% CK Red Head Normal (split Black / WB)	PB GK (/ ● WB)	
	• 50% HN Red Headed Normal (split WB)	PB GK (/ WB)	
Yellow Headed Normal PB GK	• 50% CK Red Head Normal (split Yellow / WB)	PB GK (/ ● WB)	
	• 50% HN Red Head Normal (split Yellow, / WB)	PB GK (/ ● WB)	
Black Headed White Breasted WB GK	• 50% CK Red Head White Breast (split Black)	WB GK (/ ●)	
	• 50% HN Red Headed White Breasted	WB GK	
Yellow Headed White Breasted WB GK	• 50% CK Red Head White Breast (split Yellow)	WB GK (/ ●)	
	• 50% HN Red Head White Breast (split Yellow)	WB GK (/ ●)	

Hen			
Red Headed White Breasted WB GK			

X

Cock	Progeny Expected		
Red Headed Normal PB GK	• 50% CK Red Headed Normal (split WB)	PB GK (/ WB)	
	• 50% HN Red Headed Normal (split WB)	PB GK (/ WB)	
Red Headed (split White Breasted) PB GK (/ WB)	• 25% CK Red Headed Normal (split WB)	PB GK (/ WB)	
	• 25% CK Red Headed White Breasted	WB GK	
	• 25% HN Red Headed Normal (split WB)	PB GK (/ WB)	
	• 25% HN Red Headed White Breasted	WB GK	
Red Headed White Breasted WB GK	• 50% CK Red Headed White Breasted	WB GK	
	• 50% HN Red Headed White Breasted	WB GK	
Black Headed Normal PB GK	• 50% CK Red Head Normal (split Black / WB)	PB GK (/ ● WB)	
	• 50% HN Black Headed Normal (split WB)	PB GK (/ WB)	
Yellow Headed Normal PB GK	• 50% CK Red Head Normal (split Yellow / WB)	PB GK (/ ● WB)	
	• 50% HN Red Head Normal (split Yellow / WB)	PB GK (/ ● WB)	
Black Headed White Breasted WB GK	• 50% CK Red Head White Breast (split Black)	WB GK (/ ●)	
	• 50% HN Black Headed White Breasted	WB GK	
Yellow Headed White Breasted WB GK	• 50% CK Red Head White Breast (split Yellow)	WB GK (/ ●)	
	• 50% HN Red Head White Breast (split Yellow)	WB GK (/ ●)	

Gouldian Finches - Care, Breeding & Genetics

Quick Reference Table - Gouldian Finch Progeny

MUTATIONS - YELLOW BACK

With over 50 different cocks and 40 different hens across all the mutations and split genes, there are approximately 2,000 possible combinations, so only the most popular combinations can be included here. Consult an online Gouldian calculators for any other pairings.

Cock		
Red Headed Yellow Back (Euro)	PB YB	

X

Hen		Progeny Expected		
Red Headed Normal	PB GK	50% CK / 50% HN	Red Headed Dilute (SF YB) / Red Headed Normal	PB DK (/ YB) / PB GK
Red Headed Yellow Back	PB YB	50% CK / 50% HN	Red Headed Yellow Back DF / Red Headed Yellow Back	PB YB / PB YB
Black Headed Yellow Back	PB YB	50% CK / 50% HN	Red Head Yellow Bk DF (split Black) / Red Headed Yellow Back	PB YB (/ ●) / PB YB
Yellow Headed Yellow Back	PB YB	25% CK / 25% HN	Red Head Yellow Bk DF (split Yellow) / Red Head Yellow Back (split Yellow)	PB YB (/ ●) / PB YB (/ ●)
Red Head White Breast Yellow Back	WB YB	50% CK / 50% HN	Red Head Yellow Back DF (split WB) / Red Headed Yellow Back (split WB)	PB YB (/ WB) / PB YB (/ WB)
Black Headed Normal	PB GK	50% CK / 50% HN	Red Head Dilute (SF YB) (split Black) / Red Headed Yellow Back	PB DK (/ ●) / PB YB
Yellow Headed Normal	PB GK	50% CK / 50% HN	Red Head Dilute (SF YB) (split Yellow) / Red Head Yellow Back (split Yellow)	PB DK (/ ●) / PB YB (/ ●)

Hen		
Red Headed Yellow Back (Euro)	PB YB	

X

Cock		Progeny Expected		
Red Headed Normal	PB GK	50% CK / 50% HN	Red Headed Dilute (SF YB) / Red Headed Normal	PB DK (/ YB) / PB GK
Red Headed (split Yellow Back SF)	PB GK (/ YB)	25% CK / 25% CK / 25% HN / 25% HN	Red Headed Yellow Back / Red Headed Dilute (SF YB) / Red Headed Yellow Back / Red Headed Normal	PB YB / PB DK (/ YB) / PB YB / PB GK
Red Headed Yellow Back	PB YB	50% CK / 50% HN	Red Headed Yellow Back / Red Headed Yellow Back	PB YB / PB YB
Red Headed White Breast (split Yellow Back SF)	WB YB (/ YB)	25% CK / 25% CK / 25% HN / 25% HN	Red Headed Yellow Back (split WB) / Red Headed Dilute (SF YB) (split WB) / Red Headed Yellow Back (split WB) / Red Headed Normal (split WB)	PB YB (/ WB) / PB DK (/ YB WB) / PB YB (/ WB) / PB GK (/ WB)
Red Head White Breast Yellow Bk DF	WB YB	50% CK / 50% HN	Red Headed Yellow Back (split WB) / Red Headed Yellow Back (split WB)	PB YB (/ WB) / PB YB (/ WB)
Black Headed Normal	PB GK	50% CK / 50% HN	Red Head Dilute (SF YB) (split Black) / Black Headed Normal	PB DK (/ ●) / PB GK
Yellow Headed Normal	PB GK	50% CK / 50% HN	Red Head Dilute (SF YB) (split Yellow) / Red Headed Normal (split Yellow)	PB DK (/ ●) / PB GK (/ ●)

Gouldian Finches - Care, Breeding & Genetics

Genetic Considerations

Quick Reference Table - Gouldian Finch Progeny

MUTATIONS - AUSTRALIAN YELLOW

With over 50 different cocks and 40 different hens across all the mutations and split genes, there are approximately 2,000 possible combinations, so only the most popular combinations can be included here. Consult an online Gouldian calculators for any other pairings.

Cock				
Red Headed Australian Yellow	WB AY			
X				

Hen		%	Type	Progeny Expected	
Red Headed Normal	PB GK	50%	CK	Red Headed Normal (split AY)	PB GK (/ AY)
		50%	HN	Red Headed Normal (split AY)	PB GK (/ AY)
Red Headed (split Australian Yellow)	PB GK (/ AY)	25%	CK	Red Headed Normal	PB GK
		25%	CK	Red Headed Normal (split AY)	PB GK (/ AY)
		25%	HN	Red Headed Normal	PB GK
		25%	HN	Red Headed Normal (split AY)	PB GK (/ AY)
Red Headed Australian Yellow	WB AY	50%	CK	Red Headed Australian Yellow	WB AY
		50%	HN	Red Headed Australian Yellow	WB AY
Black Headed Normal	PB GK	50%	CK	Red Headed Normal (split Black, split AY)	PB GK (/ • AY)
		50%	HN	Red Headed Normal (split AY)	PB GK (/ AY)
Yellow Headed Normal	PB GK	50%	CK	Red Headed Normal (split Yellow, split AY)	PB GK (/ • AY)
		50%	HN	Red Headed Normal (split Yellow, split AY)	PB GK (/ • AY)

Hen				
Red Headed Australian Yellow	WB AY			
X				

Cock		%	Type	Progeny Expected	
Red Headed Normal	PB GK	50%	CK	Red Headed Normal (split AY)	PB GK (/ AY)
		50%	HN	Red Headed Normal (split AY)	PB GK (/ AY)
Red Headed (split Australian Yellow)	PB GK (/ AY)	25%	CK	Red Headed Normal	PB GK
		25%	CK	Red Headed Normal (split AY)	PB GK (/ AY)
		25%	HN	Red Headed Normal	PB GK
		25%	HN	Red Headed Normal (split AY)	PB GK (/ AY)
Red Headed Australian Yellow	WB AY	50%	CK	Red Headed Australian Yellow	WB AY
		50%	HN	Red Headed Australian Yellow	WB AY
Black Headed Normal	PB GK	50%	CK	Red Headed Normal (split Black / AY)	PB GK (/ • AY)
		50%	HN	Black Headed Normal (split AY)	PB GK (/ AY)
Yellow Headed Normal	PB GK	50%	CK	Red Head Normal (split Yellow / AY)	PB GK (/ • AY)
		50%	HN	Red Head Normal (split Yellow / AY)	PB GK (/ • AY)

Legend - Progeny

	Red Head	PB	Purple Breast	GK	Green Back	PK	Pastel Blue Back	• Split Black Head
	Black Head	WB	White Breast	YB	Yellow Back (EU)			○ Split Yellow Hd SF
	Yellow Head			AY	Australian Yellow		Red Tip Beak	◐◑ Split Yellow Hd DF
	Salmon Head	SF	Single Factor	SK	Silver Back		Yellow Tip Beak	/ Split (various)
	Lt Salmon Head	DF	Double Factor	DK	Dilute Back		White Tip Beak	

Gouldian Finches - Care, Breeding & Genetics

Quick Reference Table - Gouldian Finch Progeny

MUTATIONS - BLUE

With over 50 different cocks and 40 different hens across all the mutations and split genes, there are approximately 2,000 possible combinations, so only the most popular combinations can be included here. Consult an online Gouldian calculators for any other pairings.

Cock		Progeny Expected	
Red Headed Blue Backed	PB BK		

X

Hen		Progeny Expected	
Red Headed Normal	PB GK	• 50% CK Red Headed Normal (split Blue) • 50% HN Red Headed Normal (split Blue)	PB GK (/ BK) PB GK (/ BK)
Red Headed (split Blue Backed)	PB GK (/ BK)	• 25% CK Red Headed Blue • 25% CK Red Headed Normal (split Blue) • 25% HN Red Headed Blue • 25% HN Red Headed Normal (split Blue)	PB BK PB GK (/ BK) PB BK PB GK (/ BK)
Red Headed Blue Backed	PB BK	• 25% CK Red Headed Blue • 25% HN Red Headed Blue	PB BK PB BK
Black Headed Normal	PB GK	• 50% CK Red Head Normal (split Black / Blue) • 50% HN Red Headed Normal (split Blue)	PB GK (/ • BK) PB GK (/ BK)
Black Headed (split Blue Backed)	PB GK (/ BK)	• 25% CK Red Headed Blue (split Black) • 25% CK Red Head Normal (split Black / Blue) • 25% HN Red Headed Blue • 25% HN Red Headed Normal (split Blue)	PB BK PB GK (/ • BK) PB BK PB GK (/ BK)

Hen			
Red Headed Blue Backed	PB BK		

X

Cock		Progeny Expected	
Red Headed Normal	PB GK	• 50% CK Red Headed Normal (split Blue) • 50% HN Red Headed Normal (split Blue)	PB GK (/ BK) PB GK BK
Red Headed (split Blue Backed)	PB GK (/ BK)	• 25% CK Red Headed Blue • 25% CK Red Headed Normal (split Blue) • 25% HN Red Headed Blue • 25% HN Red Headed Normal (split Blue)	PB BK PB GK (/ BK) PB BK PB GK (/ BK)
Red Headed Blue Backed	PB BK	• 50% CK Red Headed Blue • 50% HN Red Headed Blue	PB BK PB BK
Red Headed Pastel Blue (SF YB)	PB PK (/ YB)	• 25% CK Red Headed Blue • 25% CK Red Headed Pastel Blue • 25% HN Red Headed Blue • 25% HN Red Headed Silver	PB BK PB PK (/ YB) PB BK WB SK (/ BK YB)
Black Headed Normal	PB GK	• 50% CK Red Headed Normal (split Black, Blue) • 50% HN Black Headed Normal (split Blue)	PB GK (/ • BK) PB GK (/ BK)
Yellow Headed Normal	PB GK	• 50% CK Red Head Normal (split Yellow, Blue) • 50% HN Red Head Normal (split Yellow, Blue)	PB GK (/ • BK) PB GK (/ • BK)

© 2025 - Hanks, Tony

Genetic Considerations

Quick Reference Table - Gouldian Finch Progeny

MUTATIONS - AUST. VARIEGATED BLUE

With over 50 different cocks and 40 different hens across all the mutations and split genes, there are approximately 2,000 possible combinations, so only the most popular combinations can be included here. Consult an online Gouldian calculators for any other pairings.

Partner # 1		
Red Headed Aust Variegated Blue	WB GB	
X		
Partner # 2	**Progeny Expected**	
Red Headed Normal — PB GK	• 50% CK Red Headed Normal (split AVB)	PB GK (/ BK AY)
	• 50% HN Red Headed Normal (split AVB)	PB GK (/ BK AY)
Red Headed (split AVB) — PB GK (/ BK AY)	• 12.5% CK Red Headed AVB	WB GB
	• 12.5% CK Red Headed Normal (split AVB)	PB GK (/ BK AY)
	• 12.5% CK Red Headed Blue (split Aust Yellow)	PB BK (/ AY)
	• 12.5% CK Red Headed Aust Yellow (split Blue)	WB AY (/ BK)
	• 12.5% HN Red Headed AVB	WB GB
	• 12.5% HN Red Headed Normal (split AVB)	PB GK (/ BK AY)
	• 12.5% HN Red Headed Blue (split Aust Yellow)	PB BK (/ AY)
	• 12.5% HN Red Headed Aust Yellow (split Blue)	WB AY (/ BK)
Red Headed Blue Backed — PB BK	• 50% CK Red Headed Blue (split Aust Yellow)	PB BK (/ AY)
	• 50% HN Red Headed Blue (split Aust Yellow)	PB BK (/ AY)
Red Headed Australian Yellow — WB AY	• 50% CK Red Headed Aust Yellow (split Blue)	WB AY (/ BK)
	• 50% HN Red Headed Aust Yellow (split Blue)	WB AY (/ BK)
Red Headed (split Blue Backed) — PB GK (/ BK)	• 25% CK Red Headed Normal (split AVB)	PB GK (/ BK AY)
	• 25% CK Red Headed Blue (split Aust Yellow)	PB BK (/ AY)
	• 25% HN Red Headed Normal (split AVB)	PB GK (/ BK AY)
	• 25% HN Red Headed Blue (split Aust Yellow)	PB BK (/ AY)

Partner # 1		
Red Headed Blue Backed — PB BK		
X		
Partner # 2	**Progeny Expected**	
Red Headed Australian Yellow — WB AY	• 50% CK Red Headed Normal (split AVB)	PB GK (/ BK AY)
	• 50% HN Red Headed Normal (split AVB)	PB GK (/ BK AY)

NOTE: Pairings are not shown as Cocks & Hens in this table because all pairings are Red-Headed and the traits of AVB, AY & Blue are all autosomal, thus inherited independently of the sex-chromosomes.

Partner # 1		
Red Headed (split Blue) — PB GK (/ BK)		
X		
Partner # 2	**Progeny Expected**	
Red Headed (split Australian Yellow) — PB GK (/ AY)	• 12.5% CK Red Headed Normal	PB GK
	• 12.5% CK Red Headed Normal (split AVB)	PB GK (/ BK AY)
	• 12.5% CK Red Head Normal (split Aust Yellow)	PB BK (/ AY)
	• 12.5% CK Red Headed Normal (split Blue)	PB BK (/ BK)
	• 12.5% HN Red Headed Normal	PB BK
	• 12.5% HN Red Headed Normal (split AVB)	PB BK (/ BK AY)
	• 12.5% HN Red Head Normal (split Aust Yellow)	PB BK (/ AY)
	• 12.5% HN Red Headed Normal (split Blue)	PB GK (/ BK)

Gouldian Finches - Care, Breeding & Genetics

Quick Reference Table - Gouldian Finch Progeny

MUTATIONS - AUST. VARIEGATED BLUE (Cont'd)

With over 50 different cocks and 40 different hens across all the mutations and split genes, there are approximately 2,000 possible combinations, so only the most popular combinations can be included here. Consult an online Gouldian calculators for any other pairings.

Parent # 1: Red Headed (split AVB) [PB GK] (/ BK AY)		
Parent # 2	**Progeny Expected**	
Red Headed (split AVB) [PB GK] (/ BK AY)	• 3.12% CK Red Headed Normal	[PB GK]
	• 3.12% CK Red Headed Blue	[PB BK]
	• 3.12% CK Red Headed Aust Yellow	[WB AY]
	• 3.12% CK Red Headed AVB	[WB GB]
	• 6.25% CK Red Headed Normal (split Blue)	[PB GK] (/ BK)
	• 6.25% CK Red Head Normal (split Aust Yellow)	[PB GK] (/ AY)
	• 12.5% CK Red Headed Normal (split AVB)	[PB GK] (/ BK AY)
	• 6.25% CK Red Headed Blue (split Aust Yellow)	[PB BK] (/ AY)
	• 6.25% CK Red Headed Aust Yellow (split Blue)	[WB AY] (/ BK)
	• 3.12% HN Red Headed Normal	[PB GK]
	• 3.12% HN Red Headed Blue	[PB BK]
	• 3.12% HN Red Headed Aust Yellow	[WB AY]
	• 3.12% HN Red Headed AVB	[WB GB]
	• 6.25% HN Red Headed Normal (split Blue)	[PB GK] (/ BK)
	• 6.25% HN Red Head Normal (split Aust Yellow)	[PB GK] (/ AY)
	• 12.5% HN Red Headed Normal (split AVB)	[PB GK] (/ BK AY)
	• 6.25% HN Red Headed Blue (split Aust Yellow)	[PB BK] (/ AY)
	• 6.25% HN Red Headed Aust Yellow (split Blue)	[WB AY] (/ BK)
Red Headed Normal [PB GK]	• 12.5% CK Red Headed Normal	[PB GK]
	• 12.5% CK Red Headed Normal (split Blue)	[PB GK] (/ BK)
	• 12.5% CK Red Head Normal (split Aust Yellow)	[PB GK] (/ AY)
	• 12.5% CK Red Headed Normal (split AVB)	[PB GK] (/ BK AY)
	• 12.5% HN Red Headed Normal	[PB GK]
	• 12.5% HN Red Headed Normal (split Blue)	[PB GK] (/ BK)
	• 12.5% HN Red Head Normal (split Aust Yellow)	[PB GK] (/ AY)
	• 12.5% HN Red Headed Normal (split AVB)	[PB GK] (/ BK AY)
Red Headed Blue Backed [PB BK]	• 12.5% CK Red Headed Normal (split Blue)	[PB GK] (/ BK)
	• 12.5% CK Red Headed Normal (split AVB)	[PB GK] (/ BK AY)
	• 12.5% CK Red Headed Blue	[PB BK]
	• 12.5% CK Red Headed Blue (split Aust Yellow)	[PB BK] (/ AY)
	• 12.5% HN Red Headed Normal (split Blue)	[PB GK] (/ BK)
	• 12.5% HN Red Headed Normal (split AVB)	[PB GK] (/ BK AY)
	• 12.5% HN Red Headed Blue	[PB BK]
	• 12.5% HN Red Headed Blue (split Aust Yellow)	[PB BK] (/ AY)

© 2025 - Hanks, Tony

The Science of Gouldian Colours

Genetic mutations in Gouldian finches lead to variations in their plumage colours, so understanding the mechanisms behind these colours is interesting for appreciating the diversity within these birds.

The colours of Gouldians are produced by a combination of different pigments. There are also "Structural Colours" due to the microscopic feather structures that can reflect and refract light in various ways. There are three main types of pigments for Gouldian colours: Melanins, carotenoids and pteridines.

Melanins

Melanins are pigments that generate colours varying from black, brown, reddish-brown to yellow. There are two types of melanins: Eumelanin produces black and dark brown colours, while pheomelanin results in reddish-brown and yellow hues. They are produced by natural synthesis, but tightly regulated by the bird's genetics.

These pigments influence the colouration of heads, chests, and backs. For example, the black-headed Gouldian has high levels of eumelanin in its head feathers, which results in a deep black colour. Melanins also contribute to the different shades of brown and grey observed in their plumage.

Carotenoids

Carotenoids are pigments acquired by birds through their diet and deposited in the birds' feathers, beaks and skin. They result in vibrant yellow, orange and red colours. The specific carotenoids are lutein and zeaxanthin (including cryptoxanthin); while canthaxanthin is synthesized from lutein within the bird. Carotenoids also have important anti-oxidant and immune functions.

For instance, the red-headed Gouldian finch has a red head because of the accumulation of carotenoids in its feathers. Similarly, the yellow-headed variety results from different

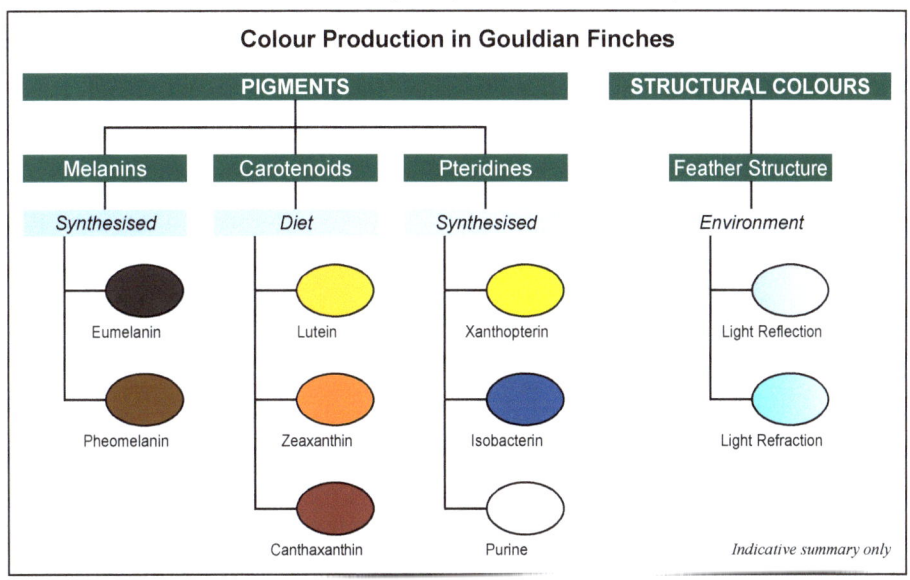

carotenoid deposition.

Pteridines

Pteridines are pigments that produce red, orange and yellow colours, similar to carotenoids, as well as producing blue colours. However, these pigments are synthesized by the birds themselves, rather than needing to be acquired through their diet. They include xanthopterin, isobacterin, and purine, as well as isoxanthopterin for fluorescence. Pteridines are less common than melanins and carotenoids, but they contribute significantly to the colouration.

In Gouldian finches, pteridines contribute to the red and orange seen in their plumage. When combined with carotenoids, they can increase the intensity of these colours, giving the birds a more vivid appearance.

Genetic Influences on Colouration

The coloration of Gouldian finches is influenced by a complex genetic basis involving multiple genes that regulate the expression and distribution of pigments.

The red-headed morph, or version, is due to a dominant allele that deposits red carotenoids in head feathers. The black-headed morph comes from a recessive allele producing eumelanin, causing black coloration. The yellow-headed morph results from another recessive allele depositing yellow carotenoids.

The inheritance of these head colour follows the principles of Gregor Mendel, with each occurring in a predictable pattern described earlier in this chapter.

Gene mutations can change the melanin synthesis in the breast

Gouldian finches can also display colour mutations as discussed in this chapter. These variations are said to be influenced by polygenic factors, because they involve the interaction of multiple genes to produce the final phenotype (appearance).

Environmental Influences on Colouration

While genetics play a significant role, three environmental factors can also influence the expression and intensity of Gouldian colours: Diet, health and sunlight.

Diet is a crucial factor in pigmentation, as the availability of carotenoids in the diet directly impacts the intensity of red and yellow colours. A diet rich in carotenoid-containing foods, such as fruits and seeds, can enhance the vibrancy of the plumage. Conversely, a diet lacking in these nutrients will result in duller colours.

Health and well-being can also influence colouration. Healthy birds typically exhibit brighter and more vivid colours, as their bodies are better able to produce and deposit pigments. In contrast, birds that are stressed, malnourished, or suffering from illness

Genetic Considerations

may have less vibrant plumage.

Sunlight exposure also affects the colouration of Gouldians. This is because sunlight helps in the synthesis of certain pigments and can enhance the overall brightness of the plumage. Birds that receive ample sunlight often display more intense colours compared to those kept in low-light conditions, while inadequate sunlight can lead to the distorted pigments seen in the condition known as Melanism. This is discussed under "Excessive Melanin" on page 78, in the chapter about "Health & Wellness".

Pigment Changes in Gouldian Mutations

The mutations described in this chapter are all related to different colours expressed in the phenotypes, or appearances. Those mutations occur because the genetic changes either enhance, reduce or block different pigments, or alter the structure of the feathers themselves. For example:

- White-Breasted .. Removal of pheomelanin only in the breast feathers, due to mutations in the genes that regulate melanin synthesis. There is also a reduced impact of carotenoids in this area.
- Lilac-Breasted .. Reduced pheomelanin production in all feathers of the body, again due to mutation changing the regulation of melanin synthesis.
- Dilute .. Reduced eumelanin production resulting in lighter colours.
- Yellow Back .. This mutation causes reduced melanins and enhanced deposition of yellow carotenoids in the feathers, overshadowing other pigments and creating a uniform yellow appearance on its back.
- Australian Yellow .. Due to reduced melanin and changes in carotenoid metabolism, the intensity of the yellow coloration is directly linked to the concentration and deposition of carotenoids, lutein and zeaxanthin, within the feathers.
- Blue .. This colour is partly caused by reduced yellow carotenoids - lutein, zeaxanthin and cryptoxanthin. However, it is not produced by pigments alone, but also by Structural Colours created by the scattering of light due to the nanostructures of keratin in the feathers. Because of the structural changes causing light interference, where specific wavelengths are

White-Breasted

Dilute (SF Yellow Back)

Yellow Back

Australian Yellow

Blue

Cinnamon

Seagreen

Mutation pigment results

amplified, the wavelengths corresponding to the blue spectrum are enhanced, resulting in the vivid blue colour.
- Cinnamon . . This mutation is another example with reduced melanin production. Eumelanin is significantly reduced, as demonstrated again by the grey appearance of the head of the Black-Headed version of Cinnamon birds.
- Seagreen . . This mutation alters the expression of genes responsible for the synthesis, deposition and balance of carotenoids and melanins in the feathers, as well as a subtle iridescence.

Essentially, all mutations work by changing the balance and interactions between feather pigments, as well as the microscopic feather structure with resulting light interference. Ref

COMMON CHALLENGES & SOLUTIONS

This book has addressed many aspects associated with the care of Gouldians. The following subjects are highlighted as common challenges.

Recognizing and Addressing Breeding Issues
Chick Tossing

Chick tossing is a concerning and frustrating behaviour that is sometimes observed with Gouldian finches. This is where one of the parents ejects their chicks from the nest before they are ready to fledge – usually resulting in death by starvation or cold temperatures.

The behaviour can be triggered by various factors:

- Environmental stress
- Inadequate nutrition
- Inexperienced parent birds (too young)
- Health problems in the chicks
- Inadequate nesting materials
- Insects in the nest
- Intrusive nest inspections
- Disturbances by other birds.

If chick-tossing occurs it is crucial for breeders to intervene promptly to prevent harm to the chicks.

Solutions are to address the causes listed above as well as using nests that are designed to make chick-tossing less likely; adjusting the diet; ensuring a calm and secure environment; and removing any meddlesome companions or independent juveniles.

The bird most likely to be doing the chick-tossing is the cock when he wants to start the mating cycle again. If problems continue it may be the best solution to remove the male from the breeding cabinet so the hen can raise the chicks in peace.

Excessive Heat

Overheating of nests during the breeding season can create challenges for Gouldian finches. Transparent polycarbonate roofing in many aviaries or bird rooms, along with high nest positions favoured by birds, can lead to elevated temperatures near the ceiling, causing excessive heat in the nest box.

Consistent temperatures above 40°C can affect the incubation process, potentially leading to lower hatchability and increased deaths of chicks. Eggs may develop incorrectly or simply not hatch at all. Even when chicks do hatch under high-temperature conditions, they may be weaker and more susceptible to health issues.

Roof insulation above nests to block excessive heat

Breeders should ensure that the breeding environment avoids extreme temperatures. This can be achieved through roof insulation above nests; proper ventilation; providing shade; and using cooling systems such as extraction fans or air conditioning units. Additionally, offering sufficient fresh water and maintaining a balanced diet can help birds manage heat stress effectively.

Health Problems and Treatments

Air Sac Mites

Air sac mite infestations are a common and serious health concern for Gouldian finches. Air sac mites (Sternostoma tracheacolum) are microscopic parasites that infest the respiratory system, particularly the air sacs, trachea and bronchi. These mites cause significant distress and may be fatal if not treated promptly.

Infestations are transmitted from native birds that perch on the aviary wire and subsequently spread rapidly among the birds within the Gouldian flock.

Symptoms of an air sac mite infestation include laboured breathing, beak open when perched, tail bobbing, wheezing, clicking sounds when breathing, and a general decline in the bird's activity and appetite. In severe cases, the mites can lead to respiratory failure and death.

To diagnose an air sac mite infestation, breeders should look for symptoms or consult a veterinarian who can examine respiratory secretions or use an endoscope to visualize the mites. Holding the bird close to your ear may reveal a regular clicking sound when it breathes. In severe cases, this clicking can be heard from the flight or cabinet at night.

The treatment of air sac mite infestations requires the administration of specific anti-parasitic medications, such as ivermectin. This can be done either orally via water or topically by applying a drop to the bird's skin. A three-day course administered through water is highly effective; however, it must be repeated after two weeks to eliminate newly hatched mites that were in the egg stage during the initial treatment.

To prevent mite re-infestation it is important to treat all birds within the aviary, not only those exhibiting symptoms. In addition, a comprehensive cleaning and disinfection of the enclosure and equipment are crucial to eradicate any remaining mites and their eggs.

Preventive measures include maintaining hygiene in the aviary, Diatomaceous Earth, avoiding overcrowding, regularly checking birds for signs of illness, and routine preventative use of Ivermectin if outbreaks are common. Providing an environment with minimal stress and a balanced diet can also support the birds' immune system, reducing susceptibility to infestations.

Injuries Due to Predators

Predators can be a risk to Gouldian finches housed in outdoor flights. Common predators include cats, rats, snakes and other birds, which may cause injuries or fatalities. Among native birds, Butcherbirds and Miner Birds are known to pose particular challenges.

Protecting the aviary with proper wiring and conducting regular monitoring are essential measures. However, Gouldians may still experience injuries due to fright and panic, even without direct physical contact.

Butcherbirds are especially persistent

An effective additional protection measure is installing wind breaks that also prevent wild birds from landing on the wire surface of the flight. Some people achieve this with a shade-cloth blind outside the wire, while others use fixed transparent shields left ajar for ventilation. With an area narrowed to less than 150mm, wild birds are unwilling to enter the space.

Vitamin D3 Deficiency

Vitamin D3 deficiency is a common issue for Gouldians in flights and breeding cabinets, often due to insufficient exposure to natural sunlight. This deficiency can result in weak bones, fragile feather quality, reduced immunity and poor breeding outcomes.

- Causes:

    The use of modern building materials for aviaries, flights and bird rooms often leads to inadequate natural sunlight. While these materials are used to protect the birds, they can result in a deficiency of UV light and Vitamin D. For instance, corrugated polycarbonate roofing is commonly used to provide light, but it blocks 100% of UV transmission. Although UV-transmitting materials were available in the past, they have been banned in most countries due to concerns about their potential to cause skin cancers in humans.

- Treatments:

    Birds in outdoor environments will naturally obtain adequate UV light. However, for those kept indoors, it is necessary to provide a source of UV light; or to supplement their diet; or more commonly do both of these things.

    Birdkeepers can use specialized UV lamps designed for birds, or "Growth Lights" marketed for indoor plants. These provide more wavelengths, but it is still necessary to include vitamin D supplements in their diet.

Iodine Deficiency in Feather Loss

The trace element iodine may be another cause of feather loss, as it is important for the health and well-being of aviary birds.

Beyond nutritional needs, iodine is connected to physiological functions such as feather development and maintenance. So feather loss observed in aviary Gouldians can sometimes be linked to an iodine deficiency.

Example of feather loss due to a vitamin or iodine deficiency

The role of iodine is that it is required for the production of thyroid hormones, including thyroxine (T4) and triiodothyronine (T3). These hormones regulate various metabolic processes, such as growth, development and energy expenditure. Sufficient iodine levels support the proper functioning of the thyroid gland, which in turn affects feather growth and the moulting cycles.

The thyroid gland produces hormones essential for feather formation. Iodine deficiency disrupts thyroid function, reducing T4 and T3 production. This imbalance can then cause delayed moulting, abnormal feather structure, or feather loss.

- Deficiencies:

    Several factors can lead to iodine deficiency in aviary birds like Gouldians. One significant cause is insufficient dietary intake of iodine, particularly if their diet relies heavily commercial bird feed without adequate supplementation, or lacks iodine-rich foods like certain seeds and vegetables.

    There can also be environmental factors that play a role, such as low iodine levels in water or soil; consumption of certain plants that interfere with iodine uptake; or gastrointestinal disorders that impair iodine absorption.

- Symptoms:

    Feather loss can indicate iodine deficiency, along with other symptoms. Feathers may appear brittle, dull, prone to breakage and moulting may be delayed. There

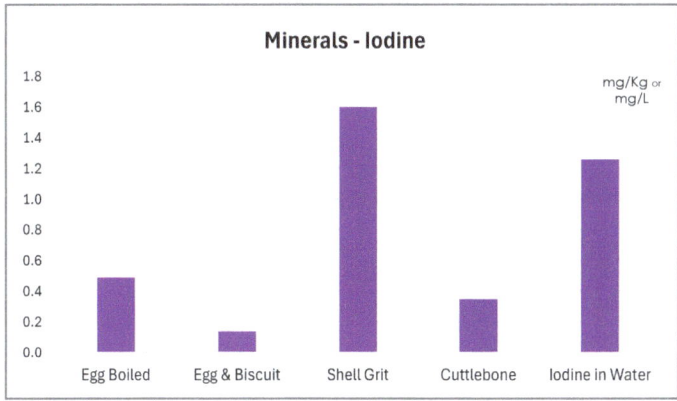

can also be swelling in the neck region, known as goitre, due to an enlarged thyroid gland caused by iodine deficiency. This can serve as a critical diagnostic marker.

- Behaviour:

  Gouldian finches with iodine deficiency may exhibit behaviours such as reduced activity and lethargy, along with a decreased appetite and weight loss.

- Treatment:

  The most effective approach to treating iodine deficiency is through prevention via dietary supplements. Foods such as eggs, shell grit and fortified bird feed can supply the required levels of iodine.

  However, iodine is frequently added into commercial aviary supplements, or as an alternative provided by a veterinarian. Examples include Morning Bird "Liquid Iodine" - a product that includes the claim of "eliminates balding in Gouldian finches". Another good source of iodine is Vetafarm "Soluvite D". There are also options that allow the birds to self-regulate their iodine intake, such as Probird "Calcium & Iodine Bells", or Avico "Iodine and Calcium Grit Buttons".

Iodine button, drops for drinking water, or blended grit

See "References Used in This Book" on page 236.

Behavioural Concerns and Solutions
Territorial Disputes
Gouldian finches typically gather in flocks and are generally peaceful birds. However, territorial disputes can occur, especially in confined or overcrowded environments. These disputes may result in aggression, feather plucking and potential injuries if not properly managed.

To prevent aggressive behaviour, it is essential to ensure sufficient space for each bird within the aviary or flight. A recommended guideline is to provide a minimum of 0.3 square meters (M2) of area per bird to avoid overcrowding.

Adding more perches and foliage can reduce stress, while multiple feeding and watering stations will help to minimize competition and aggression during feeding times. Another effective strategy is to ensure that the birds in a flight are mixed for age and balanced in terms of gender.

After careful observation of the flock it may also be necessary to separate the aggressive bird temporarily, or reintroduce it later after a period of isolation.

After observing the flock, it may be necessary to permanently relocate the aggressive bird, or temporarily separate it and reintroduce it after a period of isolation.

Abnormal Head Movements
Both Twirling and Star-gazing are abnormal behaviours observed in Gouldian finches, but they have different causes and should be differentiated. Twirling is associated with illness or injury, while Star-Gazing is related to environmental factors or nutrition.

Twirling involves the bird spinning in circles, twisting their head to an extreme angle, and occasionally falling off the perch in severe cases. Birds with this condition exhibit impaired motor functions and may fall over when they are on the ground.

Twirling

Possible causes include:
- Nutritional deficiency associated with specific colour mutations. For example, twirling may occur in Blue Gouldians due to low levels of Vitamin A. This mutation is blue becasue they are unable to absorb and process carotenoids. So they also do not obtain the antioxidants needed for Vitamin A production in their bodies.
- Neurological problem caused by Avian Paramyxovirus (PMV), for which no cure is currently known. Death results from starvation, but if the bird lives long enough they may eventually shed the virus after approximately 8 weeks. During

this period, PMV transmits readily so isolation is necessary to protect other birds.
- Head trauma, potentially from "night fright";
- Neurological problems due to excessive inbreeding;
- A fungal infection;
- A foreign body in the ear, such as a seed husk;
- Ear mites; or an ear infection.

To eliminate these possibilities, treatment with Vitamin supplements, Nystatin, Moxidectin and antibiotics are recommended. Electrolytes should also be added to the bird's drinking water to support recovery from twirling.

Star Gazing involves the bird tilting its head back and looking upwards or backwards for extended periods (torticollis).

The causes can be environmental or nutritional:

- Common environmental factors include insufficient exposure to natural sunlight, housing conditions that cause the bird to feel threatened by predators, or respiratory issues that lead birds to adopt this posture while struggling to breathe.
- Nutritional causes often stem from a poor diet, particularly a deficiency in Vitamin E.

Unfortunately, once this behaviour is established, it often becomes habitual for some affected birds.

The primary cause is environmental stress resulting from inadequate space in smaller cages, overcrowding, or a sense of confinement. Similar issues arise when perches are positioned too high, leaving insufficient room above the bird's head.

In addition to insufficient space, star-gazing behaviour is observed in birds experiencing increased fear of predators. This occurs when flights and cabinets have high gloss reflective surfaces on the ceiling, causing reflections that can trigger predator anxiety in the birds. For similar reasons, Gouldian finches should not be housed in cages that open on the ceiling, as this may cause them to frequently look upwards.

Star gazing

The treatment of star-gazing involves several steps to enhance the environment. These include moving the bird to a larger enclosure, relieving any concerns about predators, providing hiding spaces such as stress perches or foliage, and improving diet and nutrition.

It is important to provide Gouldian finches with a balanced diet that includes essential vitamins and minerals, especially Vitamin E. There are many possible causes of stargazing, including potential genetic predispositions, so affected birds are typically excluded from breeding programs. Using UV light sources and supplements can help prevent vitamin D deficiency, which may also be a contributing factor.

Poor Breeding Results

Several factors can contribute to poor breeding outcomes in Gouldian finches. A significant factor is environmental stress where insufficient space within their enclosures can result in feelings of confinement and discomfort. Small cages, overcrowding, and improper perch placement can create a stressful environment, thereby inhibiting their natural breeding behaviours.

Nutritional deficiencies can also impact breeding success and harm finches' reproductive health. Providing a balanced diet with all necessary vitamins and minerals is crucial for encouraging successful breeding.

Breeders typically separate cocks and hens during the non-breeding season. This period can be stressful for the birds due to their annual moult, but it ensures that they do not form pairs when the breeder may have other pairings planned.

Gouldian finches establish strong pair bonds during each breeding season and disruptions in these bonds can decrease their breeding motivation. It is advisable to introduce them only to potential partners that will be available to them.

When cocks and hens are placed together in a breeding flight or cabinet, they should have a two-week adjustment period before the nesting boxes are introduced.

During this period, the best possible diet is provided to simulate the natural conditions that trigger the breeding cycle. Similarly, this is also the appropriate time to introduce seeding grasses.

Additionally, genetic predispositions can impact breeding outcomes. Birds that have a history of low breeding performance or inadequate chick rearing are typically excluded from breeding programs to prevent passing on these characteristics.

Housing Concerns and Solutions
Webbing Moths in Seed Hoppers

Moths in bird seed can cause problems, especially when the seed is stored in a hopper. These insects lay eggs within the seed, leading to larvae infestation. The larvae then feed on the bird seed, contaminating it with their silk, frass (insect waste) and even their bodies, which can reduce the nutritional value of the seed.

Additionally, the silk produced by moths causes "webbing" of the seed, which disrupts the easy natural flow within the hopper. If the hopper becomes clogged and this issue is not addressed, the finches may even starve to death due to the lack of access to food.

Another problem is that the limited space within a hopper can worsen this problem, as it creates an ideal environment for the moths to thrive and multiply. Therefore it is critical

Common Challenges & Solutions

to stop moth infestations if seed hoppers are to be used.

Storing seed in airtight containers and regular cleaning the hoppers can be helpful. However, these actions will not solve the problem completely because the eggs of Flour Moths or Codling Moths are usually already in the bags of seed when purchased.

Example of webbing moths in a seed hopper

Possible solutions are as follows:

- Transferring all seed to an airtight container and storing in a fridge or freezer for 7 days will kill the moths and their eggs. However, not many people have fridge or freezer space for doing this.
- Bay Leaves are a natural deterrent, but they don't kill the moths. Many breeders attach a bay leaf in the lid of their hoppers; not in contact with the seed or where the birds can eat them.
- Garlic Powder is another natural deterrent, but there are concerns about toxicity for finches and the garlic making the seed less palatable.
- Diatomaceous Earth (DE) is an entirely safe and naturally occurring deposit derived from species of single celled diatoms. It is very effective in reducing infestations of external parasites, such as lice and mites, as well as reducing infestations of internal parasites. DE is a common feature in Shell Grit mixes and birds will even use it for a dust bath, so it is safe for them to eat. It even helps to eliminate insects in nest boxes. For moths it is sprinkled on the seed surface and mixed in. DE has a natural and physical mode of action by damaging insect shells and this will kill the moth eggs that it contacts.
- Moth Balls containing camphor as the active ingredient are effective in eliminating moths. However, it is important to ensure that they do not come into direct contact with the seeds.
- Pyrethrin or other natural insect sprays will kill flying moths, but should not be sprayed onto seed as it

Example of strips made for killing moths

will become damp.

- Clothing Moth Killers contain Transfluthrin as the active ingredient, which is a volatile synthetic pyrethroid. These products release toxic vapours that eliminate all life stages of moths. They are generally non-toxic to birds, and finch breeders have not observed any issues when using them in seed hoppers, where they effectively prevent both moth activity and webbing.

Of these, bay leaves, clothes moth killers and food-grade diatomaceous earth are recommended.

Building an Aviary: When Bigger is Not Always Better

The description of this aviary construction was originally published by the author using a pseudonym in the "Australian Birdkeeper" magazine, February-March 1992 issue.

Constructing an aviary can be a fulfilling project, however it is important to note that the size of the aviary is not the only crucial factor. Other factors such as natural sunlight and rodent-proofing are far more critical for the health and well-being of the birds.

The aviary in the diagram was constructed using quality materials, but instead of achieving the intended breeding results, it offered learning opportunities through the mistakes that were made.

The total overall size of the building was 8.4 x 6 metres (28 x 20 feet). One of the primary objectives was to protect Gouldian finches from cold draughts, which also consequently limited the amount of direct sunlight exposure.

Natural sunlight is essential for the health of birds. An aviary, regardless of its size, is inadequate if it lacks adequate exposure to natural light and does not provide artificial supplements. As discussed earlier in this book, sunlight is crucial for regulating birds' circadian rhythms, boosting their immune systems and facilitating the synthesis of vitamin D, which is vital for bone health.

Without sufficient natural light, birds can become lethargic, exhibit abnormal feather growth and have weakened immune systems. Although windows and skylights were included in this building, both glass and polycarbonate materials blocked 100% of UV light.

Another critical failure in this aviary design was the inadequate protection from mice. Rodents pose significant threats to the health of the aviary birds. They can transmit diseases, contaminate food and water supplies

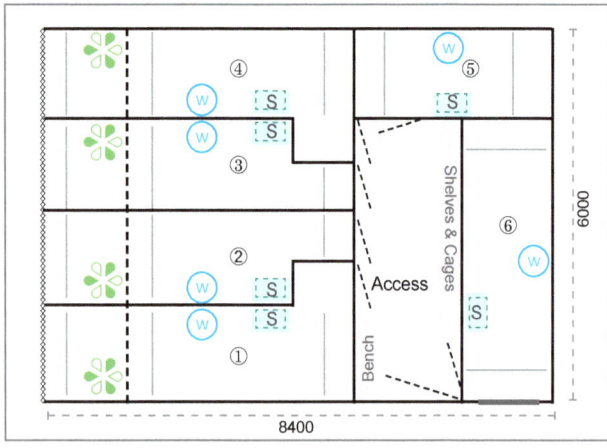

Aviary plan built in 1992 - with design flaws

and cause stress among the Gouldians.

To effectively rodent-proof an aviary, several additional measures should have been taken. The wire mesh used was the standard 12mm size (1/2 inch), which is not fine enough to prevent the entry of small mice. Another mistake was not sealing the roof with wire in addition to the roofing material itself. Barriers were installed around the base of the aviary to prevent rodents from burrowing underneath, but these barriers are ineffective if mice can enter through the wire mesh or the roof spaces.

Old 1992 photo of the aviary complex on completion

While it is tempting to think that a larger aviary provides a better environment for birds, it is crucial to balance size with functionality. A well-designed, moderately sized aviary that ensures sufficient natural light or artificial supplements, as well as protection from rodents, can provide a far superior habitat compared to a larger, poorly designed one.

Supplying Live Food

As discussed earlier under the nutrition topic of "Live Food" on page 61, this option is an excellent source of protein for Gouldian finches who learn to eat it. There are many potential live foods that are suitable, but the easiest and cleanest for most breeders is usually mealworms.

Getting Started

Mealworms, which are the larvae of the darkling beetle, can be cultivated easily. The necessary materials include plastic or glass containers with lids (such as storage bins or aquariums), bran or oatmeal as a substrate, vegetable scraps for moisture (such as carrots or potatoes), and a mesh for ventilation.

Containers must be sufficiently large to accommodate the growth of the mealworm colony. They should have smooth sides to prevent escape and be equipped with either a mesh covering or ventilation holes drilled in the lid, to allow airflow while preventing the entry of other insects.

Establishing the Colony

The container should be filled with 10 centimetres (4 inches) of clean, dry bran or oatmeal as the substrate. This material serves the dual purpose of providing both bedding and nourishment for the mealworms.

The initial batch of mealworms can be purchased and they will reproduce quickly under suitable conditions. The vegetable scraps like carrots or potatoes are added for moisture and should be replaced every 4 days.

Mealworms prefer temperatures between 21 to 27 °C (70 to 80 °F) and the substrate should be sieved and every two weeks to remove waste and dead skins.

As the mealworms eat the food they leave behind a powder called "frass". This is removed as part of the sieving process and more substrate will need to be added to replace the food that has been eaten.

Breeding Cycle

Mealworms go through four stages: egg, larva, pupa, and adult, with their life cycle lasting about 10-12 weeks. Adult beetles lay eggs that hatch into larvae, so keeping beetles in the container ensures continuous reproduction. Once there are beetles the sieved frass will contain eggs, so it should be place in a new container and not discarded.

Many people approach this by having three containers. One each beetles, pupae and larvae.

Collecting Larvae

When the mealworms grow to the desired size, they can be collected using a sieve to feed birds. Collected mealworms should be kept in a cool, dry place and storing them in a refrigerator will slow their growth to dormant and extend their life-span until they are used as bird feed.

Feeding Methods

Mealworms can be provided to birds in a shallow dish or placed on the ground to promote foraging behaviour. During training, they can also be combined with other foods to increase familiarity.

Mealworms ready to be eaten

QUESTIONS & ANSWERS

Frequently Asked Questions
Some questions have come up often, so the following list can be a useful reference . .

GENERAL

- **Sexing Birds**

    Q: What is the gender of this bird - a cock or a hen?

    A: While the normal wild-type Gouldians are easy to sex because the cock's chests are purple and the hens are mauve, this is not so easy for the various mutations. For example, a Dilute cock can look a lot like a normal hen. When making the decision the easiest things to look for are stronger colour shades on cocks, a bluer nape, a longer tail and singing or courting displays. A blackened beak is also an indication of a hen in breeding condition.

HOUSING

- **Cage Floor**

    Q: Is it suitable to use a cage or cabinet with a wire floor?

    A: No, there are too many problems with floor injuries when finches are forced to walk on a wire floor. The idea is good for droppings to fall through and keep the environment cleaner, but the injury risks are too great. Experienced Gouldian breeders line these floors with paper, or replace them.

- **Cage Size**

    Q: How many Gouldians can I keep successfully in a particular sized cage?

    A: Tall cages are not especially relevant because birds mainly fly in a horizontal direction. Therefore the height of the cage is generally ignored when calculating cage capacity. Multiply the two horizontal measurements, width and depth, to get the area of the cage. Then the capacity is one pair of Gouldians per 0.32M². (See "Guideline for Number of Birds" on page 32).

- **Housed Alone**

    Q: I had a pair of Gouldians but the hen has died. Is it alright to keep the cock in a bird-cage on his own?

    A: No, this is not recommended for two reasons. It is not a good idea to keep Gouldians in a standard bird-cage because they are too draughty. It is also not appropriate to keep them as individual pets, because Gouldians are flocking birds who need to live with their own species. If you want to avoid causing stress to your bird you should find him a new companion.

- **Predators**

    Q: Do I need to be worried about predators when designing an aviary or the position of a cage?

    A: Yes, predators need to be taken very seriously. A bird cage can never be left unattended outside - and they are unsuitable for Gouldians in any case. Possible predators include cats, but especially other birds like Mynahs, Crows and Butcherbirds. Bird wire should be 6mm (1/4 inch) to protect the birds, or double wired with a separation of 100mm (4 inches).

    In addition, Gouldian finches are very predator-aware so they are likely to be stressed or panic when threatened by a predator. This can be even worse at night with the possibility of injuries due to "night-fright".

- **Cold Temperatures**

    Q: We live in a cold climate where night-time winter temperatures can be as low as 12C (54F). Do I need to provide heating for Gouldians at night?

    A: In the wild Gouldians endure winter temperatures as low as 9C (48F), but 12C may still be too cold in your situation. In their native habitat the wind from the Australian desert drops at night. However, if you have a cold draught combined with low temperatures this can be deadly for Gouldians in captivity. It is recommended to use blinds or other barriers to protect from draughts and a safe form of heating, like a sealed oil column heater, if temperatures are going to be consistently below 14C (57F).

FEEDING

- **Boiled Eggs**

    Q: Are boiled eggs a normal part of the diet for Gouldian finches?

    A: No, eggs are not routinely fed to Gouldians because too much would be bad for them and there is also a risk of contamination. The only time that some breeders feed boiled eggs is when the finches are feeding young. In that case they can serve as a protein boost, but they are not required.

- **Boiled Eggs - Safety Time**

    Q: How long can boiled eggs be left in a bird enclosure and still be safe?

    A: It is correct to be concerned about possible food poisoning if hard boiled eggs are left un-refrigerated for too long. The safe period is affected by the temperature, but a good guide is no more than 3 hours.

- **Insects**

    Q: Do Gouldians need to be fed insects or some other type of live food?

    A: No, Gouldians do not normally eat insects. They are seed eaters and live food is not usually eaten.

- **Seed Bags - Opening**

    Q: Is there an easy way to open the stitching on 20kg seed bags?

    A: Yes, you do not need to cut the hessian bag, or untangle the stitching. When viewing the bag from the back, there are two strings in the top left-hand corner. Simply cut these and pull the main string to unravel the stitching.

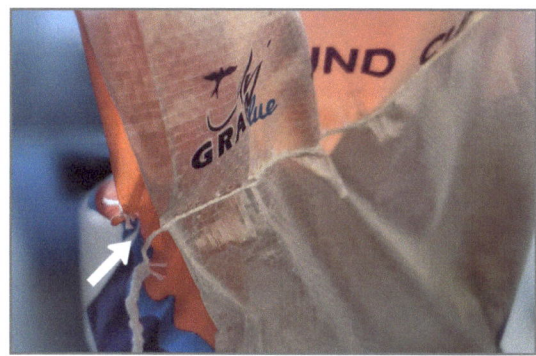

The arrow marks the cutting location on seed bags

- **Soaked Seed**

    Q: Is soaked seed something that should be provided for Gouldians?

    A: Soaked seed, sprouted seed, or chitted seed offers extra nutrition for Gouldians - especially when they are feeding chicks. Careful preparation is essential to avoid contamination. (See "Soaked & Sprouted Seed" on page 65).

- **Vegetables**

    Q: What vegetables should I be feeding to my Gouldian finches?

    A: Gouldians are seed eaters, but vegetables are also a small and important part of their diet. The birds can be fussy with vegetables if they have not tried them before. It is therefore a good idea to house juveniles with more experienced adults.

    Suitable vegetables to offer include kale, grated carrots, brussels sprouts,

Fresh seeding grasses are always attractive to Gouldians

cabbage, lettuce, chick weed, etc. However, to avoid diarrhoea it is best to restrict vegetables to once a week.

HEALTH

- **Deaths of New Birds**

    Q: I purchased a pair of new birds and the next morning one is already dead. What went wrong?

    A: There are several things that can go wrong when birds are moved to new housing. The first thing to consider is the transport itself? Due to their rapid metabolism and small bodies, Gouldians need to eat 6 to 8 times a day. They should never be in a transport cage without seed.

    Another possible issue is the stress sometimes called "Cage Fright". This can occur when a bird is moved to a new flight at the end of the day, without time to settle and find food before nightfall. If Gouldians are to be relocated, this should always be early in the day. Otherwise it is better to hold them in a smaller cabinet and introduce the larger enclosure on the next morning.

- **Feathers Falling Out**

    Q: Why do my birds all look "messy" with feathers falling out? Are they sick?

    A: This is probably the normal moult. This occurs every year, in the spring season and every bird will grow a new set of feathers. This is a particularly stressful period for the birds, so they should no longer be breeding and they should be receiving an optimal diet.

    If the feather loss is not at the time of the normal moult in Spring, it may be due to a deficiency of Vitamin D or a lack of Iodine in their diet. See "Vitamin D3 Deficiency" on page 169 and "Iodine Deficiency in Feather Loss" on page 170 for more information.

Questions & Answers

- **Head Twisting**

    Q: Why are some of my birds twisting their heads into strange positions?

    A: This is likely to be Star-Gazing, but it could also be Twirling. Star-Gazing is a normal activity in reaction to the environment and stress. Twirling is an abnormal sign of a neurological disorder or disease. See the chapter about "Common Challenges & Solutions".

- **Injuries Caused by Other Birds in the Same Cage**

    Q: What can be done about a Gouldian that had their wings damaged by a Budgerigar and now cannot fly?

    A: Firstly, Gouldian finches should never be housed with budgies because they are too dominant and are likely to hurt more delicate species. A bird with damaged wings will not survive in an aviary or flight unless there is access to both food and water at floor level.

    It is often not possible to put food and water at floor level, so the best solution is to move the injured non-flying Gouldian to a cabinet. Then provide a perch that slopes from the floor at the back of the cage to a higher position on the cage front. This will enable them to perch normally until their feathers eventually grow back or they go through the annual moult - a period that will be several months.

- **Soft Shell Eggs**

    Q: My hen is laying eggs with soft shells?

    A: Soft shells (or no shells) on eggs are a sign of calcium shortage in the diet. Calcium should be supplied as cuttlebone, shell grit and baked eggshells. However, it may also be necessary to add a vitamin supplement that includes calcium.

BREEDING

- **Breeding All Cocks**

    Q: Why are my Gouldians only producing cocks, with no hens among the offspring?

    A: There are two main reasons that this can happen. One is discussed under the heading "Offspring Survival Rates & Gender Mixes" on page 115. The other reason can be if the hen is getting too old for breeding? With birds, it is the presence or absence of the W chromosome from the female that determines the sex of the offspring. Older hens will produce less eggs and have a bias towards more males.

- **Breeding Period**

    Q: How many months of the year should I allow my Gouldians to breed?

A: In nature the normal breeding season is for the wet season when food is in abundance. In the Southern Hemisphere this is from February to July, or from August to January in the Northern Hemisphere. Breeding involves a big effort, so this period should not be extended and never during the annual moulting period.

- **Chick Tossing**

    Q: Why are my Gouldians throwing chicks out of the nest?

    A: This frustrating situation can occur for several reasons. Sometimes a sick chick is thrown from the nest as an instinct to protect the others from disease. On other occasions it is simply an accident where a chick was stuck to a parent as it left the nest. A more serious reason is when the cock wants to clean out the nest and start the breeding process again.

    If a chick has fallen out by accident it can be returned to the nest if still alive. If the tossing is a deliberate action by the cock, he can be relocated and the hen will usually raise the chicks on her own.

- **Fighting Pairs**

    Q: Is it normal that the cock and hen seem to be fighting with one another?

    A: It is not unusual for a pair of Gouldians to chase one another. This is usually because the cock is ready to breed, or sometimes because one of the birds is neglecting their share of time in the nest with eggs or chicks.

- **Nests Available**

    Q: When should nests be made available? Is there a time when they should not be available?

    A: Nests should not be available all year round. In nature Gouldians breed when food is ample, between late summer and mid-winter. It is too big a burden on the hen to breed for longer than 6 months each year, or to produce more than 2 or 3 clutches. They should also not be breeding while they are moulting.

- **Nest Inspections**

    Q: How often should I be checking to nests to see what's happening?

    A: Not very often and never if one of the parents are currently in the nest. Too many disturbances can lead the breeding pair to abandon the nest, even if there are nestlings. Most Gouldian breeders only inspect nests attached to breeding cabinets and then only if they confirm that neither parent is currently in the nest.

- **Undersized Chicks - Can They Survive**

    Q: One chick hatched several days after the rest of the clutch. It is much smaller, so can it survive?

A: No, it is unlikely that it will survive. It is a competitive environment for food in the nest and none of the larger chicks are at all concerned about their smaller sibling. Nature literally operates on the principle of "Survival of the Fittest". Unless you want to hand rear this smaller chick it is unlikely to survive.

Survival of the fittest

GENETICS

• Yellow Backed Pairings

Q: Some of my Gouldians are the Yellow Back mutation and I also have some Australian Yellows. Does it matter if I keep all of my yellow backed birds together and allow them to interbreed?

A: This is not a good idea. The genetics of the Australian Yellow and the Yellow Back are entirely different, so if you do this you will be diluting your bloodlines and causing future confusion for yourself about the possible split genes in each bird.

Yellow Back is sex-linked dominant and Australian Yellow is autosomal recessive. So any mixed pairing will produce Yellow Back phenotypes (appearances) that are all split for Australian Yellow in their genotypes, but none will express Australian Yellow.

MISCELLANEOUS

• Escaped Birds

Q: I accidentally left the door of my aviary unlatched and 5 flew away. I have put seed out to attract them back, but after 2 days I haven't seen them again. Will they survive?

A: No, sadly your Gouldian finches are not likely to survive in the wild. They are multi-generational aviary bred and they are used to being very well looked after with all of their needs met. They will struggle to find suitable food, water and shelter; then there will be the issue of predators. We'd like to think they could survive, but that's simply not probable.

• Remote Monitoring

Q: Sometimes I have to be away from home for work, so a friend feeds my birds. Is there an easy way that I can monitor things so I know they are fine?

A: Technology offers us all sorts of advantages in caring from our birds - like automatic watering systems, lighting and ventilation fans. An excellent option for monitoring a bird room or aviary is a video doorbell or indoor camera connected to your home wifi network. Brands like "Ring", "Wyse" and "Google Nest" can be quite inexpensive for the peace of mind they provide.

- **Winnowing Seed**

    Q: I have seen ads for winnowers and wondered if they work as advertised and also whether they are a good idea for saving money by reducing the amount of wasted seed?

    A: Yes, a winnower certainly does work. The basic principle is that a vacuum cleaner is used to separate the lighter weight husks and debris from the heavier whole seeds.

    If you have just one aviary, winnowing can be a good idea to separate the spilled uneaten seed and then offer it again. However, if you have multiple flights or cages, a winnower is not recommended because there may be a residue of droppings in the used seed and this creates a risk that any diseases may be spread between enclosures.

    There is also the issue of the dust created by the winnowing process. This may contain pathogens that could be harmful to humans when inhaled - especially Salmonella bacteria for example. For this reason a dust-mask is recommended to be worn when operating a winnower.

Conclusion

Caring for Gouldian finches can be a rewarding experience. Their beauty and charm bring joy to any household, and with proper care and attention, they can thrive and bring even more delight to their carers.

Recap of 12 Essential Care and Breeding Tips

1. **Proper Housing** - Provide adequate space for breeding pairs by preventing overcrowding. Ensure there is enough horizontal space for flying and room for the nest box. Make sure the environment is free of draughts, secure and protected from predators.

2. **Ideal Nesting Conditions** - Gouldian finches prefer enclosed nest boxes. For colonies in flights provide several nesting sites to reduce competition. Line the boxes with soft materials like coconut fibres or dried grass; and supply extra materials.

3. **A Balanced Diet** - Offer a well-balanced diet that includes high-quality seed mix, fresh vegetables, sources of calcium and supplements, especially during the breeding season.

4. **A Clean Water Supply** - Ensure that fresh, clean water is provided daily for both drinking and bathing purposes. Adequate hydration is essential for maintaining the health of adult finches as well as their chicks.

5. **Fail-Safe Protections** - When planning to be absent for over 24 hours, it is advisable to install "fail-safe" systems that offer secondary independent sources of food and water. Ensure that no aspect is left to chance.

6. **Use Genetics to Plan Breeding** - Understand genetics to predict the possible progeny of all potential pairings. Use this information to avoid a situation where a normal and split bird cannot be identified.

7. **Avoid Inbreeding** - Carefully select breeding pairs based on their health and genetic diversity. Inbreeding should be avoided to prevent genetic defects, reduce the likelihood of weaker birds, and to lower mortality rates.

8. **No Unplanned Breeding** - Try to avoid placing Gouldians into pairings where preferred traits will be dominated by others. Where possible combine matching head colours and avoid mixing mutations where one will eliminate the other. For example, do not mix Yellow Back and Australian Yellow.

9. **Avoid Fostering by Other Species** - In the long term it is better if Gouldians are not fostered by Bengalese finches. Chicks raised in this way will imprint on different species and ultimately be less successful when breeding themselves.

10. **Ensure Climate Protection** - Never expose Gouldian finches to conditions with cold winds. Maintain optimal temperature and humidity levels ideally between 20-30°C (68-86°F) with moderate humidity of around 60%.

11. **Reduce Stress** - Reduce disturbances around the breeding area. Provide privacy for the pairs during their breeding process. Otherwise, too much human interaction or observation will disrupt their natural behaviours.

12. **Follow an Annual Plan** - The best results will always be achieved when a plan is followed. Using the information in this book, an example of an Annual Plan is shown on page 205.

Gouldians are naturally curious

Encouragement and Support for Finch Enthusiasts

Join a Finch Club

In most countries there are bird clubs and avicultural societies for those people interested in the care and breeding of birds, finches and Gouldians. These are worthwhile to join because it's a real opportunity to learn from others with a common interest and those who have more experience.

Involvement in these groups often also leads to the opportunity to show your birds and this can bring another enjoyable aspect to the hobby of breeding Gouldians.

Join an Online Discussion Group

There are many online discussion websites that concentrate of finches, or Gouldians in particular:

- Aussie Finch Forum . . . https://www.aussiefinchforum.net/

Other discussion groups can also be found by using Google for a current search in different regions around the world.

Join a Facebook Group

One of the biggest advantages of Facebook is the availability of groups discussing specific topics. There are many that concentrate on Gouldian finches and they are very effective at getting feedback when questions arise:

- Australian Gouldian Finch
- Gouldian Finches
- Gouldian Finches of Australia
- House of Gouldians
- Lady Gouldian Finch

Alternatively, on the "Groups" page in Facebook simply search for 'gouldian' to find the latest and most active groups on the subject.

Resources for Further Learning

For more detailed information, consulting with avian veterinarians, joining bird clubs, and exploring literature on Gouldian finches can

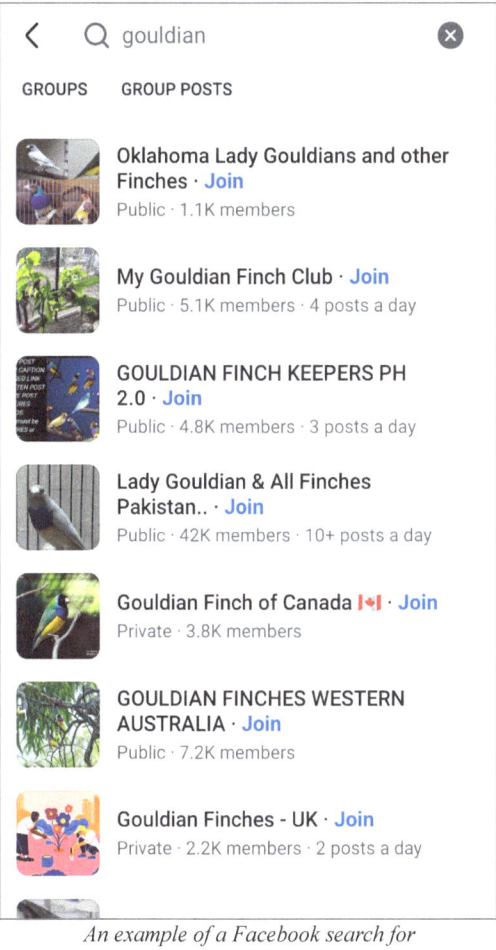

An example of a Facebook search for "Gouldian" groups

enhance your knowledge and care practices.

See "Recommended Reading" and "Recommended Resources" in the "Appendices" of this book. See page 232.

Glossary of Terms

Breeding Gouldian finches as a hobby can be both rewarding and educational. Understanding the terminology associated with bird breeding is essential for all breeders. This glossary defines common terms to help enthusiasts navigate aviculture.

Definitions & Meanings

Air Sacs	Part of the respiratory system. Can be prone to infestation by Air Sac Mites.
Allele	Genetics term for the specific variant of the gene where two copies are inherited from the two parents and can be the same or different.
Altricial	Chick hatched helpless requiring significant parental care.
Antioxidants	Substances that can protect cells from damage caused by unstable molecules called free radicals.
Apple Cider Vinegar	Vinegar made from fermented apple juice and used as a natural remedy for health problems.
Aspergillosis	Respiratory infection in birds caused by the fungus Aspergillus.
Assortative Mating	Where animals mainly select mates with similar characteristics. With Gouldian Finches this applies to head colours for example.
Australian Dilute	See Australian Recessive Dilute.
Aust Recessive Dilute	Recessive autosomal mutation characterised by muted colours (developed in Australia). Compare to the sex-linked Dilute.
Aust Variegated Blue	Secondary mutation for Gouldian finches, created by DF Australian Yellow and DF Blue in the genotype. (AVB)
Australian Yellow	Gouldian mutation characterised by a yellow body and white breast, with recessive autosomal inheritance. (Developed in Australia).
Autosomes	The chromosomes that are non-sex chromosomes.
Autosomal	Relating to any chromosome other than the sea chromosomes..
Aviary	One or more flights that provide a large enclosed space for birds; allowing space to fly and socialize.

Aviculture	The hobby of keeping and breeding birds in captivity. Helping to maintain the species in nature, preserving habitat and providing optimal care.
AVG	See Aust Variegated Blue.
AY	See Australian Yellow
Beak	The hard, pointed part of a bird's mouth used for eating and grooming.
Belly	Lower part of the front of the bird.
Bib	Patch of feathers on the throat, often contrasting with the breast and belly. Also called Chin.
Bird Claw Trimmers	A specialty tool made for the routine trimming of claws.
Blue Backed	Gouldian recessive mutation where yellow is not expressed, so the normal green back becomes blue.
Breast	Upper part of the front of the bird – the chest.
Breeding season	The time of year when birds naturally mate and reproduce.
Brood	A group of young birds hatched at the same time from the same nest.
Brooding	The act of a bird sitting on eggs to incubate them.
Bruno	See Cinnamon.
Cabinet	An enclosed space designed for breeding one pair of birds.
Cage Fright	The stress created when a bird is moved to different housing without sufficient time to adapt.
Cage Front	Manufactured wire panels that can be fitted to the front of a breeding cabinet.
Carotenoids	Pigments acquired by birds through their diet, resulting in vibrant yellow, orange and red colours. See lutein, zeaxanthin and antioxidants.
Chicks	Feathered young birds who have not yet left the nest.
Chick Tossing	Behavioural problem where nestlings or chicks are thrown from the nest and left to die. Common when nests are disturbed, or when the cock wants to breed again.
Chin	Patch of feathers on the breast or throat, often contrasting with the breast and belly. Also called Bib.
Chitted Seeds	Seeds that are undergoing the natural process of germination, where shoots are only just starting to appear. (See Sprouted Seeds)..
Chromosomes	Nucleic acids and protein found in the cell nucleus carrying genetic information in the form of a chain of genes.
Cinnamon	Gouldian mutation with a paler browner body and sex-linked recessive inheritance.
Claws	The sharp, curved nails on a bird's feet used for grasping and climbing.
Clicked Breathing	Breathing that includes click noises and coughing. Common with air sac mite infestations and respiratory diseases.
Cloaca	The single posterior opening in birds for the common exit of the digestive, urinary and reproductive tracts (faeces, urine, eggs/sperm).
Clutch	The total number of eggs laid by a female bird during a single nesting period.

Glossary of Terms

Coccidiosis	Parasitic disease that affects the intestinal tract of birds.
Cock	Male bird.
Co-dominant	Genetic inheritance where the wild-type and mutation genes express themselves equally.
Collar	See Nape.
Combined Mutations	Mutations occurring at the same time, but not changing the appearance of one another. For example, Blue Backed and White Breasted.
Coop Cup	Small stainless-steel cup suitable for supplying food or supplements.
Corona	See Nape.
Crop	Pouch located at the base of the oesophagus, serving as a temporary storage area for food before being sent to the stomach.
Crown	Top of the head of the bird.
Cryptoxanthin	A carotenoid pigment acquired by birds through their diet.
Cuttlebone	Skeleton of cuttlefish, rich in calcium
Dead-in-Shell	A fertilised egg with a fetus that failed to hatch. Common with bacterial problems, inadequate diet or disturbances that cause incubation to be abandoned.
DF	See Double-Factor or Homozygous.
DF Pastel	See Yellow Back.
Diarrhoea	Watery droppings that occur in birds drinking large amounts of water. Common with Coccidiosis, infections and stress.
Diatomaceous Earth	A naturally occurring deposit that is very effective at reducing infestations of insects and is also a common feature in Shell Grit mixes.
Dilute	Gouldian mutation characterised by muted colours. This is a single factor Yellow Backed cock. Compare to the Australian Recessive Dilute.
Dilute Blue	Secondary Gouldian mutation resulting from both Blue and Australian Recessive Dilute.
Dilute Gouldians	Two genetically different Gouldian mutations have diluted colours. See Dilute & Aust Recessive Dilute.
Dimorphic	Species where there are noticeable differences in appearance of the two sexes - such as for wild-type Gouldian finches.
Dirt Tray	A tray to collect bird droppings and facilitate cleaning.
DNA Sexing	A method used to determine the sex of a bird by analysing its DNA, typically from a feather or blood sample.
Dominant	A gene that will express itself fully whether single-factor or double-factor (Heterozygous or Homozygous).
Dominant birds	Birds in a colony that have priority access to resources such as food and nesting sites.
Double-Factor	Two copies of the gene are present. (Also known as DF or Homozygous).
Double-Split	A genotype that includes splits for two different recessive traits. For example, split Blue and split Australian Yellow.
Draught Shield	Protective panel to block cold draughts from entering an aviary or flight.
Droppings	The mix of faeces and urine produced by birds.

Edstrom Valve	A valve designed to supply gravity fed water to birds or small animals.
Egg Binding	Serious condition in which an egg becomes lodged in the hen's reproductive tract and cannot be laid naturally.
Epistatic Mutation	Inheritance where one gene can mask the expression of another gene. For example, with the white breast of the Australian Yellow.
Enteritis	An inflammation of the small intestine, caused by a bacterial or viral infection, often resulting in lethargy and diarrhoea.
Eumelanin	A melanin pigments that produces black and dark brown colours.
Euro Dominant Dilute	See Dilute.
Euro Yellow Backed	See Yellow Back.
European Fallow	See Fallow.
European Pastel	See Yellow Back.
Exhaust Fan	Extraction fan to remove hot air from a room or space.
Fail Safe	A second system in case the first one should fail.
Fallow	Gouldian mutation characterised by softer, lighter hues, primarily impacting the melanin-based colours (developed in Europe).
Feathers	The covering of the bird, providing insulation and aiding in flight.
Feather Loss	Unexpected loss of feathers, especially on the head of Gouldians. Often due to Vitamin D or iodine deficiency, mite infestations or stress.
Finch Mix	A mixture of grass seeds in proportions suitable for feeding to finches.
Fledgling	A young bird that has recently left the nest (fledged) but still depending on their parents for food and protection.
Flight	An enclosed space designed to provide birds with ample room to fly
Fluffed Up	The appearance of fluffed up feathers on a sick bird.
Foraging Tray	A tray where birds can look for food in a natural way.
Genes	Genetic information in the chromosome of each cell.
Genetic Collapse	Negative consequence of inbreeding, specifically a decrease in the fitness of offspring, reduced survival and appearance of negative recessive traits.
Genotype	The genetic constitution of an individual bird. (As opposed to the Phenotype).
Gizzard	Digestive organ for processing food.
Going Light	The symptom in birds of progressive weight loss and declining condition. See Keel Bone.
Green Food	Spinach, kale, Lebanese cucumber, chickweed, broccoli.
Grow Lights	Full spectrum artificial lights to mimic natural sunlight.
Haldane's Rule	States that any negative genetic consequences from breeding mixed morphs are more likely to affect those with different sex chromosomes.
Hand-Feeding	Process of feeding orphaned baby birds by hand, usually with a syringe or spoon.
Head Twirling	See Twirling.
Hen	Female bird.

Glossary of Terms

Heterologous	A pair of differing chromosomes, like ZW for the sex chromosomes of hens.
Heterozygous	Only one copy of the gene is present. (Also known as Single-Factor or SF).
Homologous	A pair of chromosomes that are identical, like ZZ for the sex chromosomes of cocks.
Homozygous	Two copies of the gene are present. (Also known as Double-Factor or DF).
Hopper	See Seed Hopper.
Husk	The outer layer or protective shell of a seed. This is not eaten by Gouldian finches who de-husk their seed.
Imprinting	A critical period in a bird's early life when it forms attachments and learns behaviours from its parents.
Immune Function	The body's defence system against diseases and harmful substances. It identifies and neutralizes pathogens as well as abnormal cells
Inbreeding	The mating of individuals that closely related genetically - like siblings, cousins or parent and offspring. See Genetic Collapse.
Inbreeding Depression	Negative consequence of inbreeding, specifically a decrease in the fitness of offspring, reduced survival and negative traits. See Genetic Collapse.
Incomplete Dominance	See Co-dominant.
Incubation	Process of keeping eggs warm until they hatch.
Isobacterin	Pteridine pigment synthesized by the birds.
Isoxanthopterin	Pteridine pigment for fluorescence synthesized by the birds
Ivory	Secondary Gouldian mutation resulting from both Blue and Cinnamon.
Juvenile	A young bird that is independent of their parents but has not yet reached full adulthood and adult plumage.
Keel Bone	The bone in the centre of the rib-cage of a bird. This becomes prominent when weight is lost in the condition "Going Light"..
Lady Gould Finch	Alternative name for Gouldian Finch.
Leg Rings	Split or fixed rings used to identify birds. 2.5mm is the size recommended for Gouldian finches.
Lime	See Australian Recessive Dilute.
Lutein	A carotenoid pigment acquired by birds through their diet.
Line Breeding	Technique that involves mating closely related birds, such as siblings, cousins, or parent and offspring, to reinforce certain characteristics.
Locus	Genetics term for the specific location of a gene or genetic marker on a chromosome. (Plural loci).
Lutino	Gouldian mutation with a yellow body, white breast and red eyes. Inheritance is sex-linked recessive.
Mites	Tiny insect-like organisms.
Melanin	Pigments that generate colours varying from black, brown, reddish-brown to yellow. There are two types: Eumelanin & Pheomelanin.
Melanocytes	Melanin producing cells.

Mendelian Inheritance	Classic genetics focussing on traits controlled by single genes with two alleles, one of which may be dominant to the other. (Gregor Mendel).
Monomorphic	Species where the two sexes have a similar appearance.
Morph	Genetics term for the visible variations or phenotypes within a species that is caused by genetic mutations. For example, Gouldian head colours.
Moulting	The annual process of shedding old feathers and growing new ones.
Nape	Back of the head of the bird.
Nest Box	A structure provided by bird breeders to encourage nesting.
Nestlings	Newly hatched birds who have not developed feathers yet.
Night-Fright	A sudden, panicked reaction that occurs during sleep or at night, usually triggered by a disturbance.
Normal	The bird in its' natural wild-type colour; a purple breasted green back Gouldian.
Open-Mouth Breathing	A sign of breathing distress in Gouldians.
Outcrossing	The breeding of individuals with no known recent common ancestors, introducing new, unrelated genetic material into a breeding line.
Pair Bonding	The formation of a close and long-lasting relationship between two birds, typically for the purpose of breeding.
Papillae	Small protruding nodules located on the inside of the mouths of nestlings.
Paramyxovirus	Group of viruses that can cause diseases in birds and can spread quickly, leading to high rates of illness and death.
Pastel	See Yellow Back.
Pastel Blue	Secondary Gouldian mutation resulting from both Blue and single factor Yellow Back in cocks.
Pastel Body	See Pastel Blue.
Pastel Green	See Dilute.
Pecking Order	The social hierarchy within a group of birds where some are dominant.
Perch	Branch, bar or other object on which birds can rest or stand. Often used for roosting or in a bird cage.
Phenotype	The appearance or set of observable characteristics of a bird. (As opposed to the Genotype).
Pheomelanin	A melanin pigment that generates reddish-brown and yellow colours.
Plumage	The feathers covering a bird's body, varying in colour depending on sex, age and any mutations.
PMV	See Paramyxovirus.
Polygenic Factors	Genetics term for the interaction of multiple genes involved to produce the final phenotype.
Polymorphic	Species occurring in several different forms - such as the three different head colours for wild-type Gouldian finches..
Predator-Aware	Natural behaviour that is in reaction to the threat of possible predators
Preening	The time spent by a bird cleaning and arranging their feathers, ensuring insulation, waterproofing and proper flight.
Progeny	The offspring of a pairing of birds.

Glossary of Terms

Pteridines	Pigments that produce red, orange and yellow colours, similar to carotenoids, as well as producing blue colours.
Punnett Squares	A diagrammatic method for making genetic predictions for progeny.
Purine	Pteridine pigment synthesized by the birds
Quarantine	The practice of isolating new or sick birds from the rest of the flock to prevent the spread of disease.
Quick of the Claw	The softer sensitive core of the claw, containing blood vessels and nerves.
Rainbow Finch	Alternative name for Gouldian Finch.
Recessive	A gene that will only express itself when double-factor, or when single-factor only if it is a sex-linked trait and the bird is female.
Roosting	Settling of a bird in a protected area to rest or sleep.
Rump	Lower part of the back of the bird.
Satine	Secondary Gouldian mutation resulting from both Lutino and Cinnamon.
Satinet	See Satine.
Seagreen	Gouldian mutation with a blue-green body and sex-linked recessive inheritance.
Secondary Mutations	A mutation created by two or more other mutations occurring at the same time and changing the phenotype (appearance). For example, AVB.
Seed Hopper	Type of bird feeder that holds a large amount of seed and dispenses it to birds via gravity into a tray at the bottom.
Sexing	Determine if a bird is a cock or hen, male or female.
Sex-Linked	Trait inheritance that is carried in genes on sex chromosomes (ZZ & ZW).
SF	See Single-Factor or Heterozygous.
SF Pastel	See Pastel Blue.
Shell Grit	A natural, calcium-rich supplement made from crushed seashells, essential for strong bones, healthy eggshells and aiding in digestion.
Silver	Secondary Gouldian mutation resulting from both Blue and Yellow Back.
Single-Factor	Only one copy of the gene is present. (Also known as SF or Heterozygous).
Soaked Seeds	Seeds that have absorbed water, allowing them to rehydrate and expand in the stage prior to the natural process of germination.
Sprouted Seeds	Seeds that have undergone the natural process of germination, where it emerges from dormancy and starts to grow shoot.
Star-Gazing	An unnatural action involving the bird tilting its head back and looking upwards or backwards for extended periods. (See Torticollis).
Sternostoma tracheacolum	See Air Sac Mite.
Stress Perches	Perches that are made with a number of individual areas, each one for only one or two birds.
Styptic Powder	An agent that stops bleeding from light wounds (anti-haemorrhagic).
Subordinate Birds	Birds in a colony that defer to others for access to resources such as food and nesting sites.

Tail	The elongated feathers at the rear of a bird, used for balance and steering during flight.
Territoriality	The behaviour of defending a specific area against intruders, often observed during the breeding season.
Thermostat	Temperature controlled switch for a heater or fan.
Torticollis	A condition where the neck muscles cause the head to tilt up, turn or rotate to one side.
Trait	An inherited characteristic, like head or back colour.
Tray	See Dirt Tray, Foraging Tray and Wire Tray.
Twirling	An unnatural action that involves the bird spinning in circles, twisting their head to an extreme angle and occasionally falling off the perch. See also Star Gazing.
Vent	The external opening of the cloaca, a common chamber where the digestive, urinary, and reproductive tracts meet.
Vermin Wire	Wire mesh used in building aviaries to seal against mice and other vermin.
Vitiligo	Condition involving excessive depigmentation of some feathers, due to a loss of melanin producing cells.
Watery Droppings	See Diarrhoea.
Webbing	A silk like formation in seed created by the larvae of Flour Moths or Codling Moths. Webbing can block the free flow of seed if significant.
Wild-Type	The bird in its' natural normal colour; a purple breasted green back Gouldian.
Wimple	See Nape.
Wing	One of the paired appendages on a bird's body used for flight.
Winnowing	Process used to separate lighter weight husks and debris from heavier grains of seed. Either a vacuum or the wind can be used for this.
Wire Tray	A tray above the dirt tray so that bird droppings and seed husks fall through and facilitate cleaning.
Xanthopterin	Pteridine pigment synthesized by the birds.
YB	See Yellow Back.
Yellow Back	Gouldian mutation characterised by a yellow back that has dominant sex-linked inheritance. (Developed in Europe).
Yellow Backed Gouldians	There are two genetically different Gouldian mutations that have yellow backs. See Yellow Back & Australian Yellow.
Zeaxanthin	A carotenoid pigment acquired by birds through their diet.

© 2025 - Hanks, Tony

Gouldian Finches -
Care, Breeding & Genetics

Glossary of Terms

Abbreviations Used
The following abbreviations are commonly used in the care of Gouldian finches

General Abbreviations
- Avg .... Average
- c̄ .... With
- d .... Date of death
- DOB ..... Date of birth
- Max .... Maximum
- Min .... Minimum
- Ref .... Reference number
- w/o .... Without

Symbols
- = ... Equal to
- ≠ .... Not equal to
- ≡ .... Equivalent to
- ≈ .... Approximately equal to
- / .... Split to (genetics)
- > .... Greater than
- < .... Less than
- ≥ .... Greater than or equal to
- ≤ .... Less than or equal to

Body Parts
- B .... Breast
- C .... Chin
- GF .... Gouldian Finch
- H .... Head
- K .... Back
- L .... Belly
- N .... Nape
- P .... Plumage (feathers)
- T .... Tail
- TB .... Tip of Beak

Colours
- AK .... Australian Yellow Back
- AVB ...... Australian Variegated Blue
- AY .... Australian Yellow mutation
- B .... Black (re Head colour)
- B .... Blue (re non-Head colours)
- BG .... Blue Green
- BH .... Black Headed
- BK .... Blue Back
- CK .... Cinnamon Back
- D .... Dilute
- DB .... Dark Blue
- DK .... Dilute Back
- G .... Green (re Back colour)
- G .... Grey
- GK .... Green Back
- GB .... Grey Blue
- INO .... Lutino
- LB .... Light Blue
- LB .... Lilac Breast
- LG .... Light Grey
- LK .... Lutino Back
- LO .... Light Olive
- LR .... Light Red
- LS .... Light Salmon
- LY .... Light Yellow
- M .... Mauve
- OG .... Olive Green
- P .... Purple
- PB .... Purple Breast
- PK .... Pastel Blue Back
- R .... Red
- RH .... Red Headed
- RTB ...... Red Tipped Beak
- S .... Salmon
- SGK ...... Sea-Green Back
- SK .... Silver Back
- W .... White
- WB .... White Breast
- WTB ..... White Tipped Beak
- Y .... Yellow
- YB .... Yellow Back mutation (Euro)
- YH .... Yellow Headed
- YK .... Yellow Back
- YTB ...... Yellow Tipped Beak

Genetics

- ad .... Australian Recessive Dilute gene (recessive autosomal)
- ARD ..... Australian Recessive Dilute mutation
- AVB ...... Australian Variegated Blue, a secondary mutation
- ay .... Australian Yellow gene (recessive autosomal)
- b .... Black-Headed gene (recessive sex-linked)
- bk .... Blue-Backed gene (recessive autosomal)
- cn .... Cinnamon Back gene (recessive sex-linked)
- DF .... Double-Factor (trait on 2 chromosomes)
- Iv .... Ivory, a secondary mutation
- lu .... Lutino gene (recessive sex-linked)
- PB .... Pastel Blue, a secondary mutation
- R .... Red-Headed gene (dominant sex-linked)
- Sat .... Satine, a secondary mutation
- Sv .... Silver, a secondary mutation
- SF .... Single-Factor (trait on 1 chromosomes)
- sg .... Sea-Green Back gene (recessive sex-linked)
- W .... Sex chromosome in birds
- wb .... White-Breasted gene
- YB .... Yellow Back gene (co-dominant sex-linked)
- yh .... Yellow-Headed gene (recessive autosomal)
- Z .... Sex chromosome in birds
- Zb .... Black-Headed gene on the Z sex chromosome
- ZR .... Red-Headed gene on the Z sex chromosome
- ZW .... Female sex chromosomes in birds
- ZYB ...... Yellow Back gene on the Z chromosome
- ZZ .... Male sex chromosomes in birds
- / .... Split for a trait that does not appear in the phenotype

Nutrition

- RDA ...... Recommended Daily Allowance
- DE .... Diatomaceous Earth

Breeding

- C .... Chicks number
- E .... Eggs number
- F .... Fledged date
- H .... Hatched date
- LR .... Leg Ring
- MP .... Mating Pair
- NIS .... Nest Inspection Schedule

Health

- A .... Assessment (the diagnosis)
- ACV ...... Apple Cider Vinegar
- ASAP .... As soon as possible
- AV .... Avian Veterinarian
- BW .... Body Weight
- cc .... Chief complaint
- Dc .... Discontinue
- DNA ..... Deoxyribonucleic Acid (used in sexing birds)
- Ext .... External
- FB .... Foreign body
- F/u .... Follow-up
- Inf .... Inferior
- Moxi ..... Moxidectin
- NAD ..... No abnormality detected
- P .... Plan (The treatment)
- QOL ...... Quality of Life
- T .... Temperature
- URI .... Upper Respiratory Infection
- UV .... Ultra-Violet light
- UVB ..... Ultraviolet B (important for vitamin D synthesis)

© 2025 - Hanks, Tony

Glossary of Terms

Ratios in Aviculture

Aviculture is the practice of keeping and breeding birds, where several ratios are used to manage dietary and environment.

Carbohydrate Ratio ..... = Carbohydrate Content / Total Food Intake
Standard seed mixes have a Carbohydrate Ratio of approx 55%. Egg & biscuit mix is 5%; boiled eggs are 1%; while mealworms are 3% if live or 7% when dried. Gouldian finches need a balanced diet with approximately 50% carbohydrates and these are a primary source of energy, naturally provided in seeds.

Clutch Ratio .... = Number of Clutches / One Breeding Season
To protect the health of the hen, most Gouldian breeders remove nest boxes after two clutches per season; so an average Clutch Ratio of 2.0.

Fat Ratio .... = Fat Content / Total Food Intake
Standard seed mixes have a Fat Ratio of approximately 5%. Egg & biscuit mix is 6%; boiled eggs are 9%; while mealworms are 13% if live or 28% when dried. Gouldian finches need a balanced diet with approximately 5% fats and are used in the diet for energy, insulation and the absorption of fat-soluble vitamins. However, excessive fat can lead to obesity and related health issues.

Fledged Rate .... = Number of Fledglings / Number of Chicks
The Fledged Rate is the percentage of chicks successfully reared to the stage of becoming fledglings. For mature Gouldian finches over 12 months old a typical Fledged Rate is 75%.

Hatched Rate .... = Number of Chicks / Number of Eggs
The Hatched Rate is the number of chicks as a percentage of the number of eggs. For mature Gouldian finches over 12 months old a typical Hatched Rate is 80%.

Male Gender Ratio ..... = Number of Cocks / Number of Birds
In colony breeding programs this Male Gender Ratio is normally planned to be 45%. This is important to ensure successful mating and reduce conflicts. Having more hens helps in minimizing competition among cocks and also provides for any losses of hens due to the burdens of egg production.

Offspring Gender Ratio .. = Number of Male Offspring / Number of Offspring
When Gouldian finches are matched for head colour the Offspring Gender Ratio for normally around 45 to 50% males. When pairs are mismatched this same ratio has been recorded at 80%.

Offspring Per Pair . . . . . . = Number of Juveniles / Number of Pairs in 1 Season
With a Clutch Ratio of 2.0 and a goal of 4 offspring per clutch the Offspring Per Pair would be 8, however a more realistic average figure is closer to 6.

Offspring Survival Rate. . = Number of Fledglings / Number of Eggs
This ratio combines the results of both the Hatched Rate from eggs and the Fledged Rate from chicks. A typical Offspring Survival Rate is 60%.

Protein Ratio . . . . = Protein Content / Total Food Intake
Standard seed mixes have a Protein Ratio of approximately 15%. Egg & biscuit mix has a Protein Ratio of 12%; boiled eggs are 13%; while mealworms are 20% if live or 53% when dried to remove moisture. Protein is essential for the growth and maintenance of birds. Young birds and breeding pairs need higher Protein Ratios of between 20-25%, compared to adult maintenance rates of 15-18%.

Space Ratio . . . . = Total Floor Area / Number of Birds
The recommended minimum Space Ratio per bird is $0.32M^2$. So a flight of 2 x 3 M ($6M^2$) could house up to 18 Gouldian finches. Overcrowding can lead to stress, aggression and the spread of disease.

Temperature Humidity Ratio = Temperature C / Humidity %
An appropriate environment for Gouldian finches has a Temperature to Humidity Ratio of 50%, where a common daytime temperature is 30C with a humidity level of 60%.

RATIO BENCHMARKS	
Ratio	Value
Carbohydrate Ratio	50%
Clutch Ratio / yr	2.0
Fat Ratio	5%
Fledged Rate	75%
Hatched Rate	80%
Male Ratio (Cabinet)	50%
Male Ratio (Colony)	45%
Offspring Gender Ratio	47%
Offspring Per Pair / yr	6.0
Offspring Survival Rate	60%
Protein Ratio	20%
Space Ratio	$0.32M^2$
Temperature Humidity	50%

12

FORMS LIBRARY

The forms on the following pages are useful resources for the care and breeding of Gouldian finches. These pages are not covered by the author's copyright and are free to copy and use as needed.

To copy these forms onto standard paper sizes, set the photocopier magnification as follows:

PHOTOCOPYING TO MAKE FORMS			
Paper Size Name	Size Metric	Size Imperial	Magnification
These forms	228.6 x 152.4 mm	9 x 6 inches	-
US Letter	279.4 x 215.9 mm	11 x 8.5 inches	125%
International A5	210 x 148 mm	8.3 x 5.8 inches	95%
International A4	297 x 210 mm	11.7 x 8.3 inches	133%

Bird Log for Gouldian Finches

Ref Number			Leg Ring		Season Yr	
Description					Date Born	
			Sex	❏ Ck ❏ Hn		❏ Estimate
Phenotype:					Date Died	

Head		Tip of Beak	
❏ Red	❏ Grey	❏ Red	❏ White
❏ Black	❏ Lt Grey	❏ Yellow	
❏ Yellow	❏ Lt Salmon	Belly	
	❏ Salmon	❏ Yellow	❏ White
	❏ White	❏ Lt Yellow	

Purchased	
Date	
From	
Amount	$

Breast		Back	
❏ Purple	❏ Lilac	❏ Green	❏ Pastel Blue
❏ White		❏ Yellow	❏ White Silv
		❏ Yellow Mrks	❏ Seagreen
		❏ Dilute Grn	❏ AVB
		❏ Blue	❏

Sold	
Date	
To	
Amount	$

Genotype:

Sex Linked

Z	Z	W	Z	Z	Z	Z
❏ R ❏ b	❏ R ❏ b	❏ Hen	❏ YB	❏ YB	❏ cn ❏ sg	❏ cn ❏ sg

Autosomes			Genetics Note
❏ yh ❏ wh	❏ wb ❏ wb	❏ bk ❏ bk	❏ ay ❏ ay	

Date	General Notes, Pairings, Health, Medications, etc

Annual Plan for Gouldian Finches

Southern Hemisphere		ACTIVITY		Northern Hemisphere
JAN	Pre-Breed	☐ Pairs are in place for bonding period ☐ Nests installed	Pre-Breed	JUL
FEB	Breeding	☐ Introduce breeding triggers - seeding grasses and Greens & Grains ☐ Annual deep clean of the holding flights	Breeding	AUG
MAR		☐ Continue seeding grasses and Greens & Grains ☐ Introduce soaked seed ☐ First fledglings expected		SEP
APR		☐ Continue seeding grasses, Greens & Grains and soaked seed ☐ More fledglings expected ☐ Start removing independent juveniles		OCT
MAY		☐ Continue seeding grasses, Greens & Grains and soaked seed ☐ More fledglings expected ☐ Continue removing independent juveniles		NOV
JUN		☐ Continue soaked seed ☐ More fledglings expected ☐ Continue removing independent juveniles		DEC
JUL		☐ More fledglings expected ☐ Continue removing independent juveniles ☐ Start removing nests when hens have had 2 clutches		JAN
AUG		☐ More fledglings expected ☐ Continue removing independent juveniles ☐ Continue removing nests when hens have had 2 clutches		FEB
SEP	Post-Breed	☐ Separate cocks and hens ☐ All nests removed	Post-Breed	MAR
OCT	Annual Moult	☐ Continue premium finch mix and supplements during this period ☐ Annual deep clean of all breeding cabinets and nest boxes ☐ Replace all batteries in irrigation timers for automatic water systems	Annual Moult	APR
NOV		☐ Continue premium feed ☐ Move juveniles to separate flights for cocks and hens as they are sexed		MAY
DEC	Pairing	☐ Change to Austerity Seed to enhance the effect of feed improving soon ☐ Plan pairings based upon desirable traits and the progeny expected ☐ Select and place the pairs ready for the next breeding season	Pairing	JUN

Gouldian Finches - Care, Breeding & Genetics

Decision Tree for Health Diagnosis

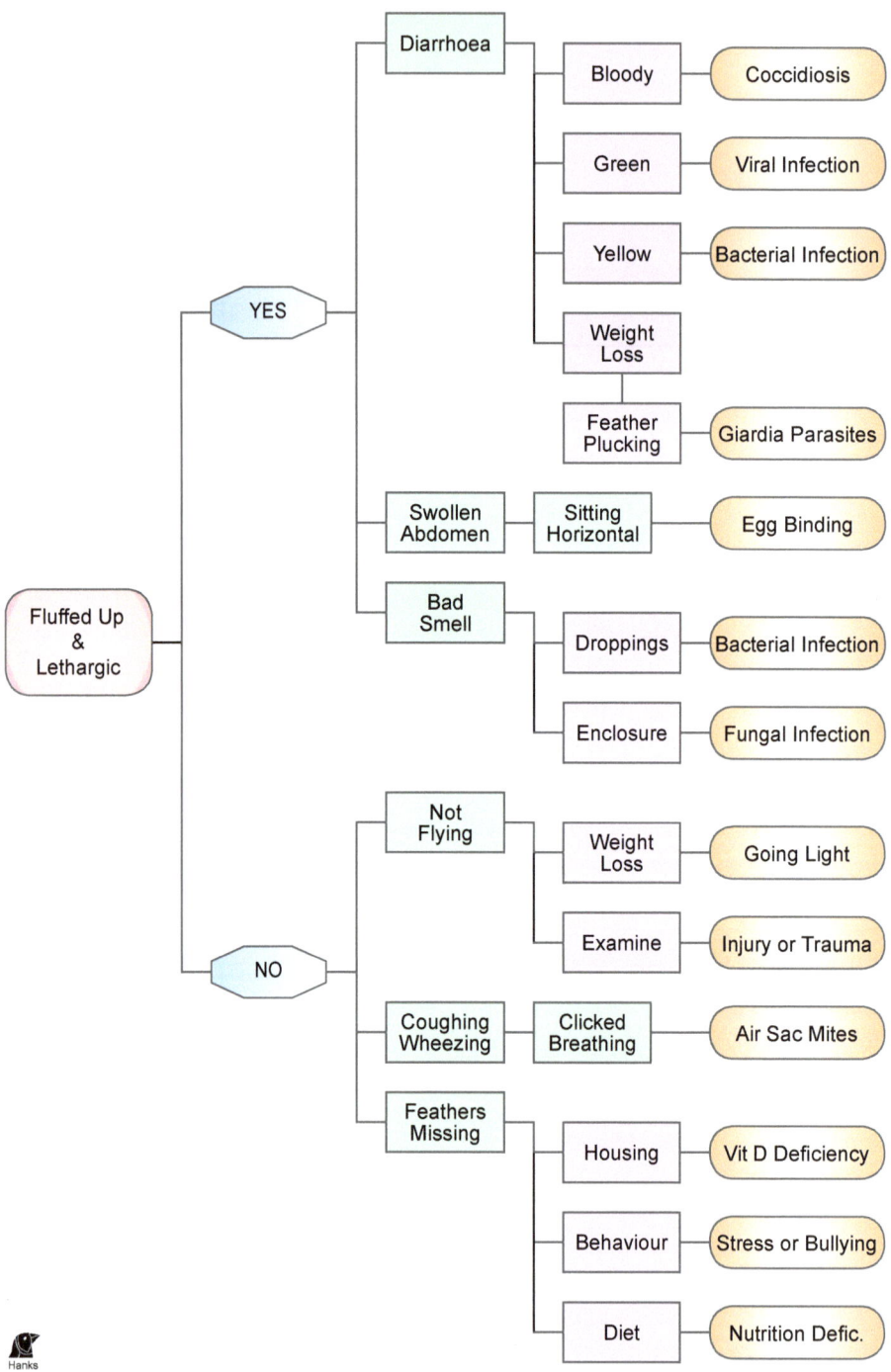

Dosage Rates

PRODUCT	BRAND	SOURCE	INDICATION	LABEL	DOSE PER	DURATION

Health Treatment Notes

Ref Number		Enclosure		Leg Ring		Treatment Start Date	
Description			❑ Cock ❑ Hen ❑ Juv				
				Location	❑ Flight ❑ Cabinet ❑ Hospital Cage		
Diagnosis:						Age:	
						Start Weight:	

Date	Treatment / Medication	Dose Rate	Duration / Notes

Result:	
	End Weight:
Learnings:	

Forms Library

Breeding Date Calculations

Incubation Started	Hatching Expected		Chicks Heard	Fledging Expected		Fledging Occurs	Independence
1	17		1	20		1	25
2	18		2	21		2	26
3	19		3	22		3	27
4	20		4	23		4	28
5	21		5	24		5	29
6	22		6	25		6	30
7	23		7	26		7	1
8	24		8	27		8	2
9	25		9	28		9	3
10	26		10	29		10	4
11	27		11	30		11	5
12	28		12	1		12	6
13	29		13	2		13	7
14	30		14	3		14	8
15	1		15	4		15	9
16	2		16	5		16	10
17	3		17	6		17	11
18	4		18	7		18	12
19	5		19	8		19	13
20	6		20	9		20	14
21	7		21	10		21	15
22	8		22	11		22	16
23	9		23	12		23	17
24	10		24	13		24	18
25	11		25	14		25	19
26	12		26	15		26	20
27	13		27	16		27	21
28	14		28	17		28	22
29	15		29	18		29	23
30	16		30	19		30	24

Red Numbers indicate the Next Month

Assumptions: *These tables are based on a 30 Day month. For 31 day months, add one day to the result. For 28 days subtract 2; 29 days subtract 1.*

Breeding Log for Gouldian Finches

Ref Nbr		Cage Nbr		Season Yr		Clutch Ref	
Description				Note			

	Bird Ref	Leg Ring	Bird Description		Date Paired	
Cock					Date Paired	
Hen					Date Nest	

Progeny Expected

- ❏ 100% Normal
- ❏ 50% Normal & 50% Split
- ❏ 100% Split
- ❏ 50% Split & 50% Mutation
- ❏ 100% Mutation
- ❏ 25% Normal, 50% Split & 25% Mutation
- ❏

	First Egg	Incubating	Hatched	Feathers	Fledged	Independent	Removed
Date							
Number							

Chicks	Leg Ring 1	Leg Ring 2	Leg Ring 3	Leg Ring 4	Leg Ring 5	Leg Ring 6	Leg Ring 7
Assessment	Nest Prep	Feed Chicks	Feed Young	Egg Binding	Leave Eggs	Leave Chicks	Chick Tossing
	/5	/5	/5	❏Y ❏N	❏Y ❏N	❏Y ❏N	❏Y ❏N

Date	General Notes, Pairings, Health, Medications, etc

Forms Library

Punnett Squares - 1 Trait - Predicted Progeny

(See "How to Use Punnett Squares" on page 146)

	COCK			
		Z	Z	
HEN	Z	Z Z	Z Z	
	W	Z W	Z W	

Progeny Expected:

	COCK			
		Z	Z	
HEN	Z	Z Z	Z Z	
	W	Z W	Z W	

Progeny Expected:

	COCK			
		Z	Z	
HEN	Z	Z Z	Z Z	
	W	Z W	Z W	

Progeny Expected:

	COCK			
		Z	Z	
HEN	Z	Z Z	Z Z	
	W	Z W	Z W	

Progeny Expected:

Punnett Squares - 2 Traits - Predicted Progeny
(See "How to Use Double Punnet Squares" on page 148)

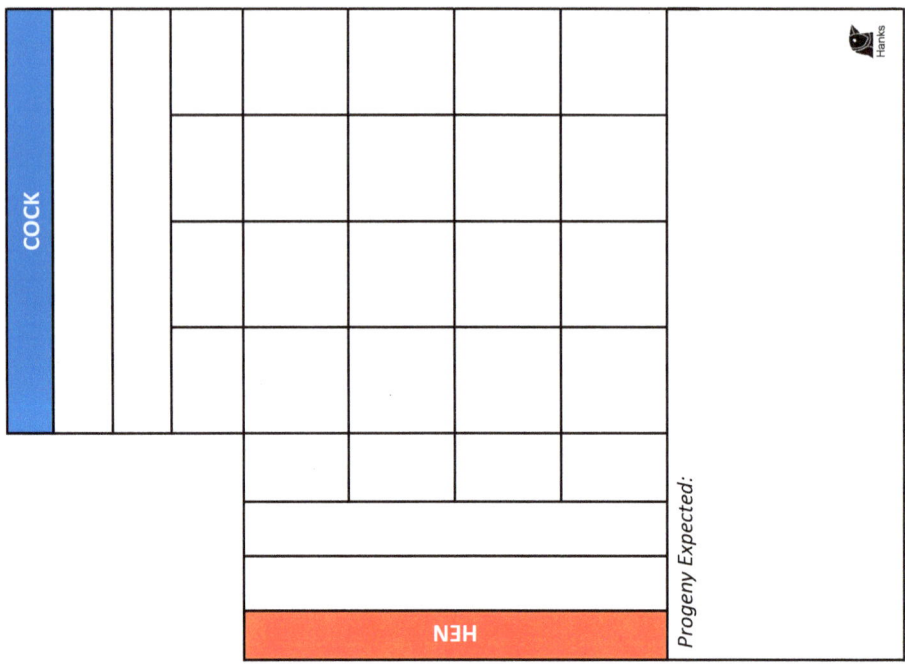

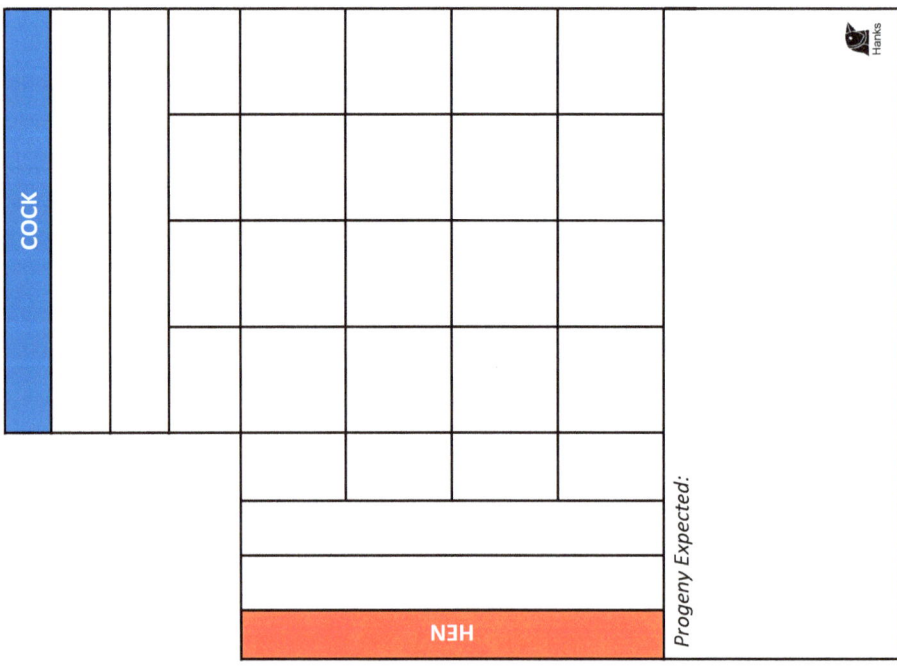

Breeding Plan Based on Predicted Progeny

Cage Nbr		Cage Nbr	
Description		Description	
Cock		Cock	
Hen		Hen	
Progeny Expected		Progeny Expected	
Cage Nbr		Cage Nbr	
Description		Description	
Cock		Cock	
Hen		Hen	
Progeny Expected		Progeny Expected	
Cage Nbr		Cage Nbr	
Description		Description	
Cock		Cock	
Hen		Hen	
Progeny Expected		Progeny Expected	
Cage Nbr		Cage Nbr	
Description		Description	
Cock		Cock	
Hen		Hen	
Progeny Expected		Progeny Expected	
Cage Nbr		Cage Nbr	
Description		Description	
Cock		Cock	
Hen		Hen	
Progeny Expected		Progeny Expected	

Offspring Log for Gouldian Finches

Ref Nbr		Cage Nbr		Season Yr		Clutch Ref	
Description				Note			

	Bird Ref	Leg Ring	Bird Description				
Cock						Date Paired	
Hen						Date Nest	

	First Egg	Incubating	Hatched	Feathers	Fledged	Independent	Removed
Date							
Number							

Chicks	Ref Nbr	Leg Ring	Gender	Body Colours	Chick Notes
1					
2					
3					
4					
5					
6					
7					
8					

Date	General Notes, Pairings, Health, Medications, etc

The gender & colours will be added later. (An alternative to the Breeding Log).

Sale Transfer Form

Date:		Leg Ring:			
Description:			Sex:	☐ Ck ☐ Hn	
Season Yr:		Date Born:		☐ Estimate	
Phenotype:	Head	Tip of Beak	Back		Belly
	☐ Red	☐ Red	☐ Green	☐ White Silv	☐ Yellow
	☐ Black	☐ Yellow	☐ Yellow	☐ Seagreen	☐ Lt Yellow
	☐ Yellow	☐ White	☐ AY Marks	☐ Cinnamon	☐ White
	☐ Grey	Breast	☐ Dilute Gm	☐ Lutino	☐
	☐ Lt Grey	☐ Purple	☐ Blue	☐ AVB	
	☐ Lt Salmon	☐ White	☐ Pastel Blu	☐	
	☐ Salmon	☐ Lilac			
	☐	☐			
Genotype:	Split . .			☐ Unknown	
Parent Ck:				☐ Unknown	
Parent Hn:				☐ Unknown	
Medic'ns:					
Comment:					
Sold By:			Ph:		Hanks

Sale Transfer Form

Date:		Leg Ring:			
Description:			Sex:	☐ Ck ☐ Hn	
Season Yr:		Date Born:		☐ Estimate	
Phenotype:	Head	Tip of Beak	Back		Belly
	☐ Red	☐ Red	☐ Green	☐ White Silv	☐ Yellow
	☐ Black	☐ Yellow	☐ Yellow	☐ Seagreen	☐ Lt Yellow
	☐ Yellow	☐ White	☐ AY Marks	☐ Cinnamon	☐ White
	☐ Grey	Breast	☐ Dilute Gm	☐ Lutino	☐
	☐ Lt Grey	☐ Purple	☐ Blue	☐ AVB	
	☐ Lt Salmon	☐ White	☐ Pastel Blu	☐	
	☐ Salmon	☐ Lilac			
	☐	☐			
Genotype:	Split . .			☐ Unknown	
Parent Ck:				☐ Unknown	
Parent Hn:				☐ Unknown	
Medic'ns:					
Comment:					
Sold By:			Ph:		Hanks

Reference Table - Gouldian Genetics Abbreviations

Foundations & Conventions:					
Sex Chromosomes		**Autosomes**		**Trait Abbreviations**	
Male / Cock	Female / Hen	(38 pair)		Dominant	Recessive
Z Z	Z W	A A		*Upper Case*	*Lower Case*
Standard Abbreviations:					
H Head	B Breast	K Back		TB Tip of Beak	C Chin N Nape
L Belly	T Tail	SF Single Factor		DF Double Factor	/ Split for
Traits on Sex Chromosomes - Z:					
R Red Headed	b Black Headed	YB Yellow Bk (Eu)		YB^{SF} Dilute (cocks only)	cn Cinnamon
sg Seagreen	ino Lutino				
Traits on Autosomes - A:					
yh Yellow Headed	PB Purple Breasted	lb Lilac Breasted		wb White Breasted	bk Blue Back
ay Aust Yellow	fa Fallow	ad Aust Recess Dilute		yh^{DF} Yellow Tip Beak (YH or BH)	bk^{DF} White Tip Beak
Secondary Mutations:					
YB^{SF} + bk Pastel Blue (cocks only)		YB + bk Silver		cn + bk Ivory	
cn + ino Satine		ay + bk Aust Variegated Blue (AVB)			

Bird Room Reference

GOULDIAN BREEDING TIMETABLE

Event	Days
First Egg After Copulation	5 Days
Incubation Period	16 Days
Hatching	0 Days
Pin Feathers *after hatching*	+6 Days
Eyes Open *	+7 Days
Full feathers *	+15 Days
Fledging *	+24 Days
Independence *	+48 Days / +24 Days *after fledging*

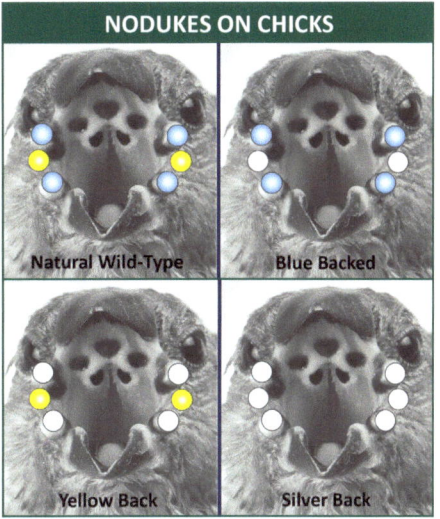

NODUKES ON CHICKS — Natural Wild-Type, Blue Backed, Yellow Back, Silver Back

Quick Reference Table - Annual Feeding Timetable

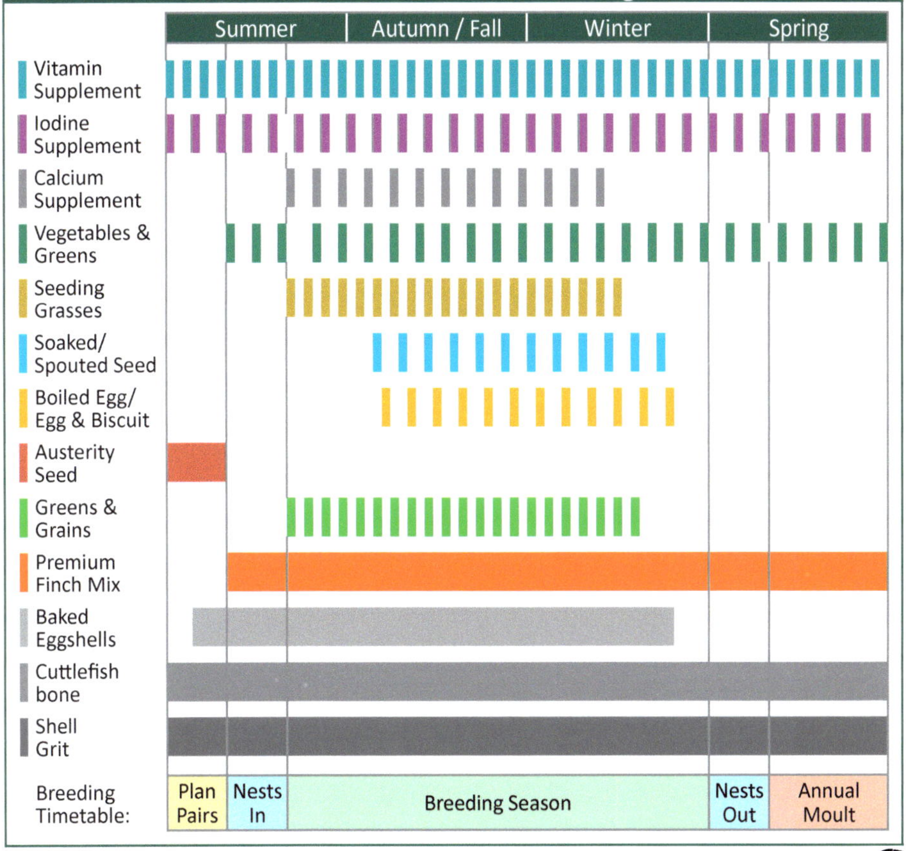

Item	Summer	Autumn / Fall	Winter	Spring
Vitamin Supplement	●	●	●	●
Iodine Supplement	●	●	●	●
Calcium Supplement		●	●	
Vegetables & Greens	●	●	●	●
Seeding Grasses		●	●	
Soaked/Spouted Seed		●	●	
Boiled Egg/Egg & Biscuit		●	●	
Austerity Seed	●			
Greens & Grains		●		
Premium Finch Mix	●	●	●	
Baked Eggshells	●	●	●	
Cuttlefish bone	●	●	●	●
Shell Grit	●	●	●	●
Breeding Timetable:	Plan Pairs / Nests In	Breeding Season	Nests Out	Annual Moult

Gouldian Finches - Care, Breeding & Genetics

Standard Care Plan for Gouldians

	❏ MAINTENANCE	❏ BREEDING / MOULTING					Date:	
	Product Name	Dosage	Days of 7	Vitamin A	Vitamin D	Vitamin E	Calcium	Iodine
Water								
Dry Seed								
Wet Food								
			Recomnd:	M5/B11000	M1/B3000	M10/B50	M2/B5000	M2.5/B2.5
			Totals:					

	Product Name	Days of 7	Product Name	Days of 7	Product Name	Days of 7
Extras	❏ Grit: _____		❏ Cuttlebone		❏ Baked Eggshells	
	❏ Greens & Grains		❏ Seeding Grasses		❏ Direct Sun >30min	
	❏ Wet Fd: _____		❏ Live Fd: _____		❏ _____	
	❏ Spinach ❏ Kale	❏ Chickwd ❏ Endive		❏ Sprouts ❏ Cucmbr	❏ _____	

	Purpose	Product Name	Days of 7	Mths of 12	Dosage	Notes
Routine Meds	❏ Coccidiosis					
	❏ Air Sac Mites					
	❏ Worms					
	❏ Nest Preparation					
	❏ Mites & Insects					
	❏ Water Disinfectant					
	❏ _____					
	❏ _____					

(All doses are per L or Kg)

GALLERY OF COLOURS & MUTATIONS

The following pages contain a photographic gallery of Gouldian finches. These images are a useful reference and they can be used as an aid to make correct identifications.

Gallery of Gouldian Colours & Mutations

ALL	"Wild Type"	Sex Linked	Autosomal	Secondary
Red Headed	Normal	Red Head	Yellow Head	Pastel Blue
Black Headed		Black Head	Purple Breast	Silver
Yellow Headed		Yellow Back	Lilac Breast	Ivory
		Dilute (SF YB)	White Breast	Satine
		Cinnamon	Blue	Aust Variegated Blue
		Seagreen	Aust Yellow	Aust Rec. Dilute Blue
		Lutino	Fallow	
			Aust Recessive Dilute	

CK Red Headed Normal

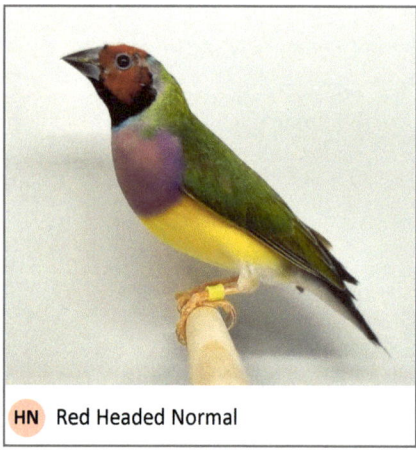

HN Red Headed Normal

CK Black Headed Normal

HN Black Headed Normal

Gallery of Gouldian Colours & Mutations

CK Yellow Headed Normal

HN Yellow Headed Normal

CK Black Headed White Breasted

HN Black Headed White Breasted

CK Red Headed Yellow Back (Euro)

HN Red Headed Yellow Back (Euro)

Gallery of Gouldian Colours & Mutations

CK Red Headed Yellow Back (Euro) Single-Factor (Dilute)

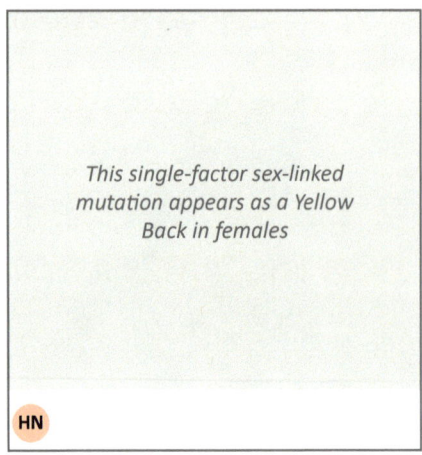

HN *This single-factor sex-linked mutation appears as a Yellow Back in females*

CK Black Headed Yellow Back (Euro) Single-Factor (Dilute)

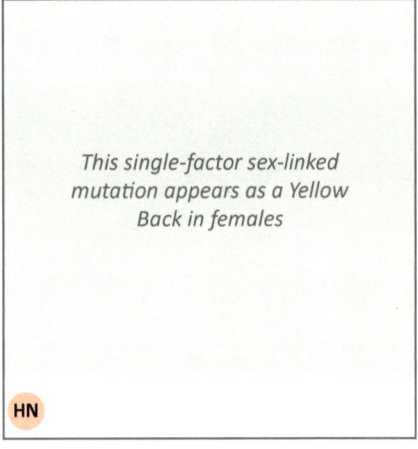

HN *This single-factor sex-linked mutation appears as a Yellow Back in females*

CK Red Headed White Breasted Yellow Back (Euro)

CK Yellow Headed White Breasted Yellow Back (Euro)

Gallery of Gouldian Colours & Mutations

CK Yellow Headed Australian Yellow (note some melanism)

HN Black Headed Australian Yellow

CK Black Headed Blue

HN Black Headed Blue

CK Red Headed Pastel Blue (SF Yellow Back & DF Blue)

This single-factor sex-linked mutation combined with DF Blue appears as a Silver in females

HN

Gallery of Gouldian Colours & Mutations

CK Red Headed Silver
(DF Yellow Back & DF Blue)

CK Black Headed White Breasted Silver
(DF Yellow Back & DF Blue)

HN Black Headed White Breasted Silver
(DF Yellow Back & DF Blue)

JV Juvenile Normal Wild-Type

JV Juvenile White Breasted

JV Juvenile Yellow Back

APPENDICES

The information in this chapter contains supporting resources for the material covered in this book, together with suggestions for further reading or study.

Nutrition Facts

	PROTEIN	FAT	CARBS	FIBRE	CALCIUM	IODINE	Refs
GRASS SEEDS:							
Finch Mix	13%	4%	62%	9%	206	-	Calc
Plain Canary	17%	4%	53%	11%	395	-	1,2,3,4,49
Fr White Millet	12%	4%	67%	6%	80	-	4,5,50,51
Panicum	11%	5%	62%	11%	220	-	4,5,52
Japanese Millet	12%	4%	67%	9%	130	-	4,53,54
Austerity Mix	14%	4%	60%	9%	237	-	Calc
Greens & Grains	11%	4%	62%	5%	187	-	4,6,7,55
Soaked Seed Mix	13%	4%	62%	9%	205	-	Calc,41
Sprouted Seed	22%	3%	53%	9%	276	-	39,40,42
OTHER PROTEIN:							
Eggs Boiled	13%	9%	1%	-	490	0.49	10-13,32
Egg & Biscuit	12%	4%	4%	1%	150	0.14	8,9
Mealworms Live	20%	13%	3%	2%	-	-	14,15
Mealworms Dried	53%	28%	7%	6%	-	-	14
MINERALS:							
Shell Grit	-	-	-	-	960,000	1.60	13
Eggshells Baked	1%	-	-	-	950,000	-	16
Cuttlebone	-	-	-	-	850,000	0.35	17,18
Iodine in Water	-	-	-	-	-	1.26	24
VEGETABLES:							
Bok Choy	2%	-	2%	1%	833	-	29,30,31
Broccoli	3%	-	5%	2%	433	-	22,32
Carrots	1%	-	9%	3%	329	-	21,32
Chickweed	5%	-	12%	7%	800	-	25,31
Endive	1%	-	3%	3%	545	-	23,31
Kale	3%	1%	5%	4%	1,613	-	19,32
Lebanese Cucumb.	1%	-	2%	1%	160	-	27,32
Lettuce	1%	-	4%	2%	280	-	28,32
Snow Pea Sprouts	6%	-	4%	2%	414	-	44,46,47
Spinach	3%	1%	3%	2%	890	0.06	20,32

Calcium & Iodine..mg per Kg or L   Calc .. Result calculated from component data.

(See "References Used in This Book" on page 236).

Vitamins & Minerals in Supplements

This table shows calculation examples for Final Delivery Concentrations after allowing for the recommended dosage, the ingredient concentration and the frequency of providing the supplement:

The formula for calculating the concentration delivered is as follows:

| Dose / Volume | X | Product Concentration | X | Days per Wk / 7 | = | Concentration Delivered |

	Birdcare Calcium	Calcium Plus	Calcivet	D Nutrical	Liquid Gold	Multi-Vite
Maker	Vetsense	Morning Bird	Vetafarm	Vetafarm	Passwell	Passwell
Format	Liquid	Liquid	Liquid	Powder	Liquid	Powder
Dosage / Volume (ml or g)	5 / 250	20 / 950	5 / 250	5 / 1000	20 / 1000	1 / 1000
Frequency (of 7 days)	7	5	7	7	7	7
VITAMIN A: [18,19] Recommended Maintenance 5,000 IU Recom Breeding or Moulting: 11,000 IU						
Ingredient (mcL or mg)						400
Ingredient (IU) [28]	0	0	0	500,000	0	1,333,333
Final Delivered (IU)	**0**	**0**	**0**	**2,500**	**0**	**1,333**
VITAMIN D3: [17,18,19,24] Recommended Maintenance 1,000 IU Recom Breeding or Moulting: 3,000 IU						
Ingredient (mcL or mg)						15
Ingredient (IU) [28]	25,000	25,000	25,000	50,000	25,000	600,000
Final Delivered (IU)	**500**	**376**	**500**	**250**	**500**	**600**
VITAMIN E: [18,20,21] Recommended Maintenance 10 IU Recom Breeding or Moulting: 50 IU						
Ingredient (mcL or mg)				500		30,000
Ingredient (IU) [28]	0	0	0	605	0	36,300
Final Delivered (IU)	**0**	**0**	**0**	**3.0**	**0**	**36.3**
CALCIUM: [17] Recommended Maintenance 2,000 mg Recom Breeding or Moulting: 5,000 mg						
Ingredient (mcL or mg)	33,000	33,000	33,000	310,000	33,000	0
Final Deliv (mcL or mg)	**660**	**496**	**660**	**1,550**	**660**	**0**
IODINE: [29] Recommended Maintenance 2.5mg Recom Breeding or Moulting: 2.5 mg						
Ingredient (mg)	0	0	0	0	0	30
Final Deliv (mg)	**0**	**0**	**0**	**0**	**0**	**0.03**
References:	8	5	2	1,16	7	11

IMPORTANT: Always check the dosage on the instructions for the actual branded product being used.

Ingredient, Recommended and Final Delivered are all per L or Kg.

Table Continued →

	Multivet	Ornithon	Solamino-vit	Soluvite D	Soluvite D Breeder	The Good Oil
Maker	Vetafarm	Inca	Allfarm	Vetafarm	Vetafarm	Wombaroo
Format	Liquid	Powder	Liquid	Powder	Powder	Liquid
Dosage / Volume (ml or g)	0.5 / 150	4 / 1000	2 / 1000	4 / 400	5 / 5000	15 / 1000
Frequency (of 7 days)	4	7	3.5	7	7	7
VITAMIN A: [18,19]	Recommended Maintenance		5,000 IU	Recom Breeding or Moulting:		11,000 IU
Ingredient (mcL or mg)		300				15,000
Ingredient (IU) [28]	94,000	1,000,000	15,000,000	1,000,000	10,000,000	50,000
Final Delivered (IU)	**179**	**4,000**	**15,000**	**10,000**	**10,000**	**750**
VITAMIN D3: [17,18,19,24]	Recommended Maintenance		1,000 IU	Recom Breeding or Moulting:		3,000 IU
Ingredient (mcL or mg)		2.5				1
Ingredient (IU) [28]	8,500	100,000	200,000	250,000	2,500,000	40,000
Final Delivered (IU)	**16**	**400**	**200**	**2,500**	**2,500**	**600**
VITAMIN E: [18,20,21]	Recommended Maintenance		10 IU	Recom Breeding or Moulting:		50 IU
Ingredient (mcL or mg)		1,920	500			2,000
Ingredient (IU) [28]	33	2,323	605	2,500	25,000	2,420
Final Delivered (IU)	**0.1**	**9.3**	**0.6**	**25.0**	**25.0**	**36.3**
CALCIUM: [17]	Recommended Maintenance		2,000 mg	Recom Breeding or Moulting:		5,000 mg
Ingredient (mcL or mg)	0	0	0	0	0	0
Final Deliv (mcL or mg)	**0**	**0**	**0**	**0**	**0**	**0**
IODINE: [29]	Recommended Maintenance		2.5mg	Recom Breeding or Moulting:		2.5 mg
Ingredient (mg)	25	0	0	200	2,000	0
Final Deliv (mg)	**0.05**	**0**	**0**	**2.0**	**2.0**	**0**
References:	1,6	9	1,3	4	27	1,10

IMPORTANT: Always check the dosage on the instructions for the actual branded product being used.

Ingredient, Recommended and Final Delivered are all per L or Kg.

(See "References Used in This Book" on page 236).

Popular Medications & Supplements

Examples Only - Always confirm instructions on product being used

PRODUCT	MAKER	INDICATION	NOTE	LABEL	DOSE RATE	DURATION	Refs
Apple Cider Vinegar	(Various)	Gut health & anti-fungal	Organic ACV	(NA)	7ml / L	1/wk, or for 3-5 days	28,29
Apple Cider Vinegar	(Various)	Anti-microbial cleaning	Organic ACV	(NA)	150ml / L	Surface cleaning	30,31
AviCalcium	VetSense	Liquid Calcium & Vit D3	Calcium 33mg/ml (3%), Vitamin D3 25,000 IU	4ml / 500ml	8ml / L	Breeding period	1,24
Avicare	Vetafarm	Disinfectant	Benzalkonium Chloride 1g / L (0.1%)	Ready-to-use	Ready-to-use	Enclosure cleaning	1,14
Aviclens	Vetafarm	Water Cleanser	Chlorhexidine Gluconate 10mg/ml (1%)	1ml / 2L	0.5ml / L	Soaking seed	1,13
Avi-Clot	Morning Bird	Blood stop styptic powder	Kaolinite 100%	Apply	Apply	As needed	1,19
Baycox Poultry	Elanco	Coccidiosis	Toltrazuril 25g/L (2.5%)	3L / 1000L	3ml / L	One dose	1,4,5
Betadine	FH Faulding	Wounds & skin infections (OTC)	Povidone-Iodine 10%	Apply	Apply	2 to 3 times/ day	1,18
Birdcare Calcium	Vetsense	Calcium supplement	Calcium 33mg/ml	5ml / 250ml	20ml / L	Ongoing	42
Calcium Plus	Morning Bird	Calcium supplement	Calcium 33mg/ml	20ml / Quart (950ml)	20ml / L	5 days / wk	43
Calcivet	Vetafarm	Calcium supplement	Calcium 33mg/ml	5ml / 250mlr	20ml / L	Ongoing	44
Chlorsig	Aspen Pharma	Eye infections (OTC)	Chloramphenicol 10 mg/g (1%)	Smear	Smear	8 hours, up to 5 days	1,17
Coccivet	Vetafarm	Coccidiosis	Amprolium 80g/L (8%), Ethopabate 5.1g/L (.5%)	1Drop / 30ml or 15ml / 10L	1.5ml / L	5-7 days	1,3
Coopex	Bayer	Residual Insecticide	Permethrin 250g/kg (25%)	25g / 5L	5g / L, or dust powder	Active for 4 weeks	1,8
D Nutrical	Vetafarm	Dietary supplement	Calcium, Vitamins & Minerals	5-20g / Kg	5g / Kg	Ongoing	1,15
DufoPlus	Bird Health	Multi-vitamin	Vitamins A, D, E and B Complex	1ml / 400ml	2.5ml / L	2 days/wk	1,34
E Powder Plus	Bird Health	Energy needs	Cultured amino acid and energy	1g / 400g	2.5g / Kg	3 days/wk	1,34
First Aid	Passwell	Emergency nutrition	Energy boost, vitamins & minerals	6g / 10ml	6g / 10ml	Prepare & feed	1,27
F-Vite Plus	Bird Health	Sterile grit & minerals	Calcium, minerals and trace elements	5g / Kg	5g / Kg	3 days/wk	1,34
Ioford	Bird Health	Health stimulant	Iodine, iron, calcium, zinc, magnesium & vit D	10ml / 2L	5ml / L	1 day/wk	1,34

IMPORTANT: These are only examples. Always check the dosage on the instructions for the actual branded product being used.

Table Continued →

Examples Only - Always confirm instructions on product being used

PRODUCT	MAKER	INDICATION	NOTE	LABEL	DOSE RATE	DURATION	Refs
KD Powder	Bird Health	Water cleanser	Acidic cleanser	1g / L	1g / L	Weekly	1,34
Liquid Gold	Passwell	Dietary supplement	Vitamin D3 & Calcium	20ml / Kg	20ml / Kg	Ongoing	45
Liquid Iodine	Morning Bird	Feather loss	Potassium Iodide 12g/L (1.2%)	2 Drops / Quart (950ml)	2 Drops / L	Ongoing	1,2
Micro-Nutrients	Naturally for Birds	Dietary supplement	Minerals and vitamins	50g / Kg	50g / Kg	Sprouted seed mix	1,39
Moxidectin	Aus Pigeon Co	Mites	Moxidectin 2mg/ml (0.2%)	5ml / L	5ml / L or spot on	24 hours, repeat 14 D	1,12
Moxivet	Pet Shop Direct	Mites	Moxidectin 1g / L (0.1%)	10ml / L	10ml / L	24 hours	7
Multi-Clens	Passwell	Water cleanser	Chlorhexidine Di-gluconate 10mg/ml (1%)	5ml /10L	0.5ml / L	Soaking seed	1,37
Multivet	Vetafarm	Dietary supplement	Vitamins, Minerals & Amino Acids	0.5ml / 150ml	3.33ml / L	Ongoing	1,16
Multi-Vite	Passwell	Dietary supplement	Vitamins & Minerals	1g /1L	1g / L	Ongoing	46
Oral Antibiotic	Aristopet	Bacterial enteritis	Tertacycline HCl 200 g / Kg (20%)	5g / L	5g / L	3-5 days	1,23
Ornithon	Inca	Dietary supplement	Vitamins & minerals	4g / L	4g / L	Ongoing	1,22
Oxymav B	Mavlab	Bacterial infections	Oxytetracycline HCl 10g/Kg	5g / 50ml	10g / 100ml	3-5 days	32,33
Prima	Naturally for Birds	Dietary supplement	Vitamins & minerals	90g / Kg	90g / Kg	Spouted seed mix	49
Prosperity	Applied Nutrition	Egg & biscuit alternative	Enzymes, vitamins & minerals	6 parts in 10	20g / 13ml	Dry or as crumble	1,40
Protein Boost	Naturally for Birds	Protein supplement	Essential amino acids & 32% protein	1 part to 4	200g / Kg	Spouted seed mix	1,38
Quik Gel	Bird Health	Energy supplement	High energy emergency gel & vitamins	5ml / 2L	2.5ml / L	As needed	1,34
Revive	Morning Bird	Anti-bacterial Anti-fungal	Pau D'Arco bark 100%	0.5g on seed	0.5g sprinkle on seed	3 days	1,25
S76	Bird Health	Mites	Ivermectin 0.8g/L & homeopathics	5ml / 2L	2.5ml / L	2 days/wk, for 4 weeks	1,34
Solaminovit Liquid	Allfarm	Dietary supplement	Vitamins & Amino Acids	2ml / L	2ml / L	Alternate days	1,10
Solaminovit Powder	Allfarm	Dietary supplement	Vitamins & Amino Acids	2g / Kg	2g / Kg in wet food 5g / Kg in dry seed	Alternate days	1,6,11

IMPORTANT: These are only examples. Always check the dosage on the instructions for the actual branded product being used.

Table Continued →

Appendices

Examples Only - Always confirm instructions on product being used

PRODUCT	MAKER	INDICATION	NOTE	LABEL	DOSE RATE	DURATION	Refs
Soluvite D	Vetafarm	Dietary supplement	Vitamins & Minerals	4g / 400ml	10g / L	Ongoing	47
Soluvite D Breeder	Vetafarm	Dietary supplmt (10x conc'n)	Vitamins & Minerals	5g / 5L	1g / L	Ongoing	48
Solvita AD3 High E	Allfarm	Dietary supplement	Vitamins A, E & D3	5ml / 500-1000g	5ml / Kg	Periodic	1
Spark	Vetafarm	Energy supplement	Glucose & electrolytes	20ml / L	20ml / L	As needed	1,9
Sulfa 3	Inca	Bacterial enteritis	Triple sulfonamide	6ml / L	6ml / L	5 - 7 days	1,21
The Good Oil	Passwell	Seed supplement	Omega 3 & 6	15ml / Kg	15ml / Kg	Breeding period	1,20
Triple C	Vetafarm	Bacterial enteritis	Chlortetracycline HCl 100mg/g. (10%)	2.5g / L	2.5g / L	5-7 days	1,26
TummyRite Plus	Applied Nutrition	Dietary supplement	Enzymes, amino acids, vitamins & minerals	50g / Kg	50g / Kg	Ongoing in dry seed	1,41
Turbo-booster	Bird Health	Health conditioner	Energy, protein and vitamins A, B, D, E & K	3ml / 500g	6ml / Kg	3 days/wk	1,34
Virkon S	Ranvet	Water cleanser	Peroxygen compounds	5g / 1L	5g / L	Soaking seed	35,36
Zade	Bird Health	Vitamin concentrate	Vitamins A, D & E	0.5ml / 2L	2.5ml / L	2 days/wk	1,34

IMPORTANT: These are only examples. Always check the dosage on the instructions for the actual branded product being used.

(See "References Used in This Book" on page 236).

Recommended Reading
- A Guide to Gouldian Finches, Their Care & Management - First Edition, ABK Publications, 1991
- A Guide to Gouldian Finches and Their Mutations - Revised Edition, ABK Publications, 2005
- Australian Birdkeeper Magazine . . . https://birdkeeper.com.au/
- Finches in Australia, Harry Doven, 2017.
- Gouldian & Finch Health, Dr Rob Marshall, 2003.
- Mendel's Laws of Inheritance . . . https://en.wikipedia.org/wiki/Mendelian_inheritance
- The Finch Keepers Recipe Book, Peter James, 2010
- The Gouldian Finch, Stewart Evans & Mike Fidler, 2005
- The Gouldian Finch Handbook, Tanya Logan, 2020

Recommended Resources
- Allfarm Animal Health . . . https://www.allfarm.com.au/
- Australian Gouldian - Don Crawford . . . www.australiangouldian.com
- Avian Directory (Health topics) . . . https://www.aviandirectory.uk/pages/illnesses.php
- Avian Microscopy . . . https://www.facebook.com/groups/146965708973398/
- Aviculture Hub . . . https://www.aviculturehub.com.au/gouldian-finch/
- Wildlife Supplies . . . https://www.wildlifesupplies.com.au/
- Bird Bands . . . https://www.birdbands.com/legrings/finch.html
- Bird Pal Avian Nutrition . . . https://birdpalproducts.com/
- Blue Gouldian Finches Blog . . . https://bluegouldians.home.blog/
- Finch Information Centre . . . http://www.finchinfo.com/birds/finches/species/lady_gouldian_finch.php
- Finch Society of Australia . . . https://www.finchsociety.org/
- Finch Stuff - Gouldian Mutations . . . https://finchstuff.com/gouldianmutations
- Gouldian Finch Calculator . . . Microsoft Store (free)
- Gouldian Finch Calculator . . . https://finchstuff.com/gouldianmutations
- Gouldian Finch Calculator . . . https://www.gouldamadinecalculator.nl/
- Gouldian Finch Society - South Africa . . . https://www.gouldianfinchsociety.co.za/
- Gouldian Finches EU, Marek Buransky . . . https://www.gouldianfinches.eu/
- Gouldian Finch.gr . . . https://gouldianfinch.gr/
- House of Gouldians - YouTube . . . https://www.youtube.com/@thehouseofgouldians
- Lady Gouldian Finch . . . https://www.ladygouldianfinch.com/
- Naturally for Birds . . . https://www.naturallyforbirds.com.au/
- Paradise Aviary - a useful collection of videos - https://www.youtube.com/@paradiseaviary
- Queensland Finch Society . . . https://qfs.org.au/
- Support for Sick Birds . . . https://www.facebook.com/groups/719419551475456/
- The Gouldian Guy . . . https://gouldianguy.com/

Products and Suppliers
In most cases international readers will need to source suppliers in their own local markets. However, these details will be a useful resource of suggested product names and models.

TIP Google is your friend. Use the product name to find a supplier in your location.

- ACP Panels . . . . . . Aluminium Composite Panels . . . . Signwriter supplies
- Aluminium Joiners External 25mm Nylon with Steel Core SPiC Engineering supplies

Appendices

- Aluminium Joiners Internal 20mm Nylon - SNCB . . . . Engineering supplies
- Aluminium Tube . . External (flights) 25x25x1.6 - 6060T5 . . . . . Metal supplies
- Aluminium Tube . . Internal (cabinets) 20x20x1.6 - 6060T5 . . . . Metal supplies
- Baycox . . . . Baycox Poultry . . . . Livestock store
- Cage Fronts . . . . . . Finch - Galvanised, coated black, 10mm sep'n . . . . Pet supplies
- Catching Net . . . . . 30cm Ring with 30cm Handle . . . . Pet supplies
- Chickweed . . . . . . Chickweed seeds . . . . Seed supplies
- Clothing Moth Killer Hovex Vaporgard Clothing Moth Killer . . . . Supermarket retailers
- Coccivet . . . . Treatment for Coccidiosis . . . . Vetafarm retailers
- Coop Cups . . . . Size "Mini" (80mm diameter) . . . . Pet supplies
- Crop Needle . . . . . Size 18 for Finches (Vetafarm) . . . . Pet supplies
- Cuttlefish Holder . . NentMent Cuttlebone Holder . . . . Amazon
- Draught Shutters . . Polycarbonate 3mm (on frame of aluminium tube) . . . . Plastics retailer
- Exhaust Fan . . . . . . Vevor Shutter Exhaust Fan 254mm EF-10-AC2 . . . . Building supplies
- FitGrit . . . . Shell grit, activated charcoal, diatomaceous earth and iodine . . . . Naturally for Birds
- Foraging Trays . . . Colour Pallet Art Trays . . . . Amazon
  Bebemoko Plastic Lab Trays . . . . Amazon
- Heater . . . . Column Oil Heater (sealed) . . . . Electrical retailer
  Cultiv8 Tube Heater 120W 61cm AQRM31003 Greenhouse supplies
  Cultiv8 Tube Heater 180W 90cm AQRM31004 Greenhouse supplies
- Iodine . . . . Morning Bird Liquid Iodine . . . . Pet supplies
- Leg Rings . . . . Numbered Split Plastic 2.5mm . . . . Pet supplies
- Lights . . . . Aimall LED Full Spectrum Grow Light Tube Strips . . . . Kogan
- Moxidectin . . . . . . Medication for air sac mites . . . . Veterinary supplies
- Mutations Posters . Gouldian Genetics - set of 5 posters . . . . . . Naturally for Birds
- Nest Boxes . . . . . . Gouldian Nesting Box . . . . Pet supplies
- Nesting Material . . Sisal & Coconut Fibre . . . . Pet supplies
- Roofing . . . . Corrugated Polycarbonate "Opal" (eg SunTuf) . . . . Building supplies
- Seed Hoppers . . . . Galvanised Steel - custom made . . . . Pet supplies
- Seed Mix . . . . Avigrain Blue "Finch Mix" . . . . Produce retailers
- Shell Grit . . . . Fine Shell Grit . . . . Pet supplies
- Snow Pea Shoots . . Show Pea Sprouting Seeds . . . . Seed supplies
- Solaminovit . . . . . . Supplement (liquid or powder) . . . . Allfarm Animal Health
- Solvita AD3 . . . . . Supplement . . . . Allfarm Animal Health
- Stress Perches . . . . Custom made in multiples of 100mm . . . . . Pet supplies
- Thermostat . . . . . . Ketotek Digital Temperature Controller . . . Amazon
- Water Baths . . . . . . ABS Resin Bird Bath for Bird Cages . . . . . . Pet supplies
- Water Controller . . Irrigation Water Timer . . . . Hardware retailer
- Water Drinkers . . . Avi One Bird Feeder Fountain 22804 . . . . . Pet Supplies
- Water Valves . . . . . Edstrom Vari-Flo Valve 1000-8000 . . . . Edstrom
- Wire . . . . Galvanised Vermin Wire (6mm) coated black. Hardware retailer

Food Preference Trials

It is important to understand the dietary preferences of Gouldian finches in an aviary for their health and conservation. Conducting a trial to determine which seeds are most preferred by birds provides useful information about their feeding habits.

Individual Seeds

Goal: The primary goal was to identify the types of seed preferred by birds in the aviary, so a variety of different seeds were selected for each food station:
- Canary Seed
- French White Millet
- Red Panicum
- Japanese Millet

Trial Design: Seed preferences vary at different times of the year, so this trial was conducted in the most important breeding season. The four food stations had equal quantities and presentation, using identical containers to avoid bias. The positions of the food stations were also rotated to eliminate location bias and a flight containing 22 Gouldian finches was selected to achieve a meaningful sample size.

Data Collected: The key observation was the total weight of the seed consumed at each food station - measured in grams to two decimal places. These results are shown in the graph below.

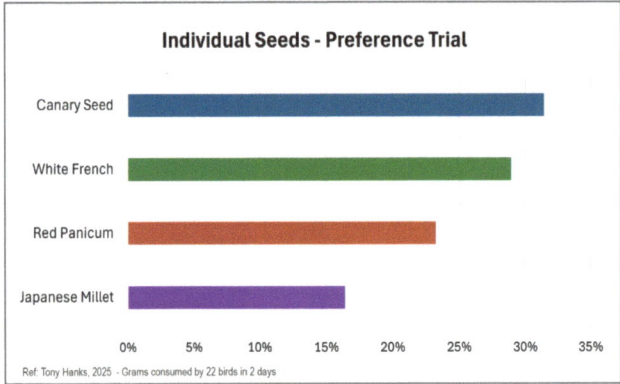

Conclusions: The most popular seeds were Canary Seed and French White Millet, then Red Panicum with Japanese Millet the least popular. This trial is the basis for the recommended mixed seed ratio in this book - see "Mixed Seed" on page 51.

The experiment will be repeated periodically to account for seasonal variations.

Appendices

Sprouted Greens

Green foods are an important part of the natural diet for Gouldian finches and new growth also offers the advantage of higher protein levels. Greens can be grown or purchased, so this trial was to measure the preferences in a commercial mix sold for salads.

Goal: The primary goal was to identify the types of sprouts preferred by birds in the aviary, so a commercial variety called a "Hero Salad" was purchased:

Broccoli Sprouts

Snowpea Sprouts

Mung Bean Sprouts

Alfalfa Sprouts

Trial Design: This trial was conducted in the most important breeding season using four food stations with equal quantities of 10 grams in identical containers to avoid bias. The positions of the food stations were also rotated to eliminate location bias and a flight containing 25 Gouldian finches was selected to achieve a meaningful sample size.

Data Collected: The key observation was the total weight of the sprouts consumed at each food station - measured in grams to two decimal places. These results are shown in the graph below.

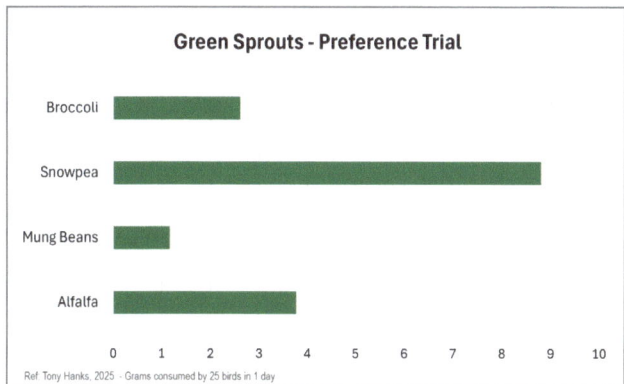

Conclusions: The most popular choice was Snowpea Sprouts by a large margin, with most of the 10g consumed.

References Used in This Book

- "Native Habitat Temperature Range" on page 17:
  - Australian Government Bureau of Meteorology - Kununurra - http://www.bom.gov.au/climate/averages/tables/cw_002038.shtml ("All available" for highest & lowest).
  - Australian Government Bureau of Meteorology - Wyndham - http://www.bom.gov.au/climate/averages/tables/cw_001013.shtml ("All available" for highest & lowest).
- "Behaviour Related to Head Colours" on page 19:
  - Pryke S: "Red dominates black: agonistic signalling among head morphs in the colour polymorphic Gouldian finch". Biological Sciences, Vol 273, Issue 1589, Page 949-957, April 2006.
  - Pryke S: "Fiery red heads: female dominance among head colour morphs in the Gouldian finch". Behavioral Ecology, Vol 18, Issue 3, Page 621-627, May 2007.
  - Williams L, King A, Mettke-Haufman C: "Colourful characters: head colour reflects personality in a social bird, the Gouldian finch, Erythrura gouldiae". Animal Behaviour, Vol 84, Issue 1, Page 159-165, July 2012.
  - Kim, KW., Jackson, B.C., Zhang, H. et al: "Genetics and evidence for balancing selection of a sex-linked colour polymorphism in a songbird". Nature Communications 10, 1852 (2019). https://doi.org/10.1038/s41467-019-09806-6.
  - https://www.aviculturehub.com.au/the-fascinating-science-of-gouldian-finch-head-colours/
  - Griffith, S: "Studies of Genetics in Australian Finch Species". White Papers, 8th International Finch Convention, QFS, 2025.
- "Examples: Vitamin D3 Concentrations in Drinking Water" on page 55:
  - https://vetafarm.com.au/product/calcivet-50ml/
  - https://www.wombaroo.com.au/product/liquid-gold/
  - https://www.wombaroo.com.au/product/multi-vite-for-birds/
  - https://www.petshopdirect.com.au/shop/item/inca-ornithon-powder-500g
  - https://chickencoach.com/products/solaminovit-liquid-200ml
  - https://www.allfarm.com.au/images/info-sheets/Info_sheet_-_Solaminovit_Liquid.pdf
- "Health Problems for Blue Gouldians" on page 91:
  - Martin, T: "Colour in Finches - Origins and Change". White Papers, 8th International Finch Convention, QFS, 2025.
- McGraw, K: "Mechanisms of Carotenoid-Based Coloring in Birds", American Union of Ornithologists, 2006.
- "Identifying Chicks in the Nest" on page 109:
  - https://www.featheredtreasuresaviary.com/identifying-mutations.html
- "Mating Preferences Related to Head Colour" on page 115:
  - Pryke S, Griffith S: "Genetic incompatibility drives sex allocation and maternal investment in a polymorphic finch". Science, Vol 323, Issue 5921, Page 1605-1607, March 2009.
  - Kim, KW., Jackson, B.C., Zhang, H. et al. Genetics and evidence for balancing selection of a sex-linked colour polymorphism in a songbird. Nature Communications 10, 1852 (2019). https://doi.org/10.1038/s41467-019-09806-6.
  - https://www.aviculturehub.com.au/the-fascinating-science-of-gouldian-finch-head-colours/
- "The Science of Gouldian Colours" on page 163:
  - Martin, T: "Colour in Finches - Origins and Change". White Papers, 8th International Finch Convention, QFS, 2025.
- "Iodine Deficiency in Feather Loss" on page 170:
  - https://morningbirdproducts.com/products/morning-bird-liquid-iodine-dietary-supplement-for-iodine-deficiencies-in-birds
- "Nutrition Facts" on page 226:
  - 1. .... https://www.alpisteseeds.com/pages/nutrients
  - 2. ... https://www.canaryseed.ca/documents/NutritionFacts
  - 3. .... https://morningbirdproducts.com/products/morning-bird-canary-seed
  - 4. .... https://www.avigrain.com.au/nutrition-information
  - 5. .... https://topflite.co.nz/
  - 6. .... https://www.nutritionvalue.org/
  - 7. .... https://www.researchgate.net/publication/318735316
  - 8. .... https://www.wombaroo.com.au/product/egg-biscuit/
  - 9. .... https://www.avione.com.au/products/bird-food2/
  - 10. .. https://www.egginfo.co.uk/egg-nutrition-and-health/
  - 11.... https://www.healthline.com/nutrition/
  - 12. .. https://www.medicalnewstoday.com/articles/283659
  - 13. .. https://www.healthdirect.gov.au/foods-

Appendices

- high-in-iodine
- 14. ... https://www.webmd.com/diet/health-benefits-mealworms
- 15. ... https://pmc.ncbi.nlm.nih.gov/articles/PMC8002850/
- 16. ... https://www.quora.com/What-is-the-nutritional-value-of-an-egg-shell
- 17. ... https://www.greatcompanions.com/product/the-benefits-of-cuttlebone-for-birds/
- 18. ... https://pmc.ncbi.nlm.nih.gov/articles/PMC8747335/
- 19. ... https://www.medicalnewstoday.com/articles/270435
- 20. ... https://www.healthline.com/nutrition/foods/spinach
- 21. ... https://www.nutritionvalue.org/Carrots
- 22. ... https://www.verywellfit.com/broccoli-nutrition-facts-calories-and-health-benefits-4118226
- 23. ... https://bivihome.com.au/product/endive/
- 24. ... https://morningbirdproducts.com/products/morning-bird-liquid-iodine-dietary-supplement-for-iodine-deficiencies-in-birds
- 25 ... https://www.mossyoak.com/our-obsession/blogs/chickweed
- 27 ... https://afcd.foodstandards.gov.au/fooddetails
- 28 ... https://www.nutritionix.com/food/lettuce
- 29 ... https://www.medicalnewstoday.com/articles/280948
- 30 ... https://time.com/3265401/14-high-calcium-foods/
- 31 ... USDA National Nutrient Database - Microsoft Copilot
- 32 ... https://fdc.nal.usda.gov/food-details/
- 39 ... https://www.researchgate.net/publication/340361936
- 40 ... https://sproutnet.com/blog/sprouts-for-optimum-nutrition/
- 41 ... https://pmc.ncbi.nlm.nih.gov/articles/PMC5332905
- 42 ... The Finch Keepers Recipe Book, James P, 2010
- 44 ... https://afcd.foodstandards.gov.au/fooddetails.aspx
- 45 ... https://www.nutritionvalue.org/Endive%2C_raw_nutritional_value.html
- 46 ... https://www.nutritionix.com/food/pea-sprouts
- 47 ... https://www.mynetdiary.com/food/calories-in-snow-pea-shoots-dau-miu-by-jade-gram
- 48 ... https://wesco.co.nz/wp-content/uploads/2016/07/
- 49 ... https://makardi.com.ua/en/canary-seeds-phalaris/
- 50 ... https://afcd.foodstandards.gov.au/fooddetails.aspx
- 51 ... https://www.nutritionvalue.org/Millet%2C_raw_nutritional_value.html
- 52 ... https://www.researchgate.net/profile/M-Kamatar/publication/272565236
- 53 ... https://twobrothersindiashop.com/blogs/food-health/barnyard-millet-benefits
- 54 ... https://slism.com/calorie/101139/
- 55 ... https://pmc.ncbi.nlm.nih.gov/articles/PMC10497464/

- "Vitamins & Minerals in Supplements" on page 227
  - 1 ..... Directions on bottle label
  - 2 ..... https://vetafarm.com.au/product/calcivet-50ml/
  - 3 ..... https://www.allfarm.com.au/products/products-type/nutrition
  - 4 ..... https://vetafarm.com.au/product/soluvite-d-25g/
  - 5 ..... https://morningbirdproducts.com/products/morning-bird-calcium-plus-liquid-calcium-formula
  - 6 ..... https://vetafarm.com.au/product/multivet-liquid-50ml/
  - 7 ..... https://www.wombaroo.com.au/product/liquid-gold/
  - 8 ..... https://www.vetsense.com.au/product/vetsense-birdcare-calcium-liquid/
  - 9 ..... https://www.petshopdirect.com.au/shop/item/inca-ornithon-powder-500g
  - 10 ... https://www.wombaroo.com.au/product/good-oil-for-birds/
  - 11.... https://www.wombaroo.com.au/product/multi-vite-for-birds/
  - 16 ... https://vetafarm.com.au/product/dnutrical-pdr-150g/
  - 17 ... https://birdsnways.com/calcium-phosphorus-vitamin-d3/
  - 18 ... https://veteriankey.com/avian-nutrition/
  - 19 ... https://www.msdvetmanual.com/management-and-nutrition/nutrition-exotic-and-zoo-animals/nutrition-in-psittacines
  - 20 ... https://www.dineachook.com.au/blog/vitamin-e-deficiency-in-chickens/
  - 21 ... https://pmc.ncbi.nlm.nih.gov/articles/PMC4463624/
  - 24 ... Naturally for Birds, Update 3 (NFBIU3) - Vitamin D3 in NFB Supplements.pdf
  - 27 ... Vetafarm - Data Sheet_Soluvite D breeder.pdf
  - 28 ... https://www.omnicalculator.com/health/mcg-to-iu-converter
  - 29 ... Klasing K: "Comparative Avian Nutrition".

Page 266-267, CABI Publishing, 1998.

- "Popular Medications & Supplements" on page 229:
  - 1 ..... Directions on bottle label
  - 2 ..... https://morningbirdproducts.com/products/morning-bird-liquid-iodine-dietary-supplement-for-iodine-deficiencies-in-birds
  - 3 ..... https://vetafarm.com.au/product/coccivet-50ml-domestic/
  - 4 ..... https://www.aussiefinchforum.net/viewtopic.php?t=23000#
  - 5 ..... https://farmanimal.elanco.com/au/poultry/product-directory/baycox
  - 6 ..... Calculation adjusted for feeding where seed husk is removed
  - 7 ..... https://pet-shop.au/shop/bird-products/chicken-products/moxidectin-wormer/
  - 8 ..... https://resources.bayer.com.au/resources/uploads/label/file12052.pdf
  - 9 ..... https://vetafarm.com.au/product/spark-liquid-conc-50ml/
  - 10 ... https://www.allfarm.com.au/images/info-sheets/Info_sheet_-_Solaminovit_Liquid.pdf
  - 11.... https://www.allfarm.com.au/images/info-sheets/Info_sheet_-_Solaminovit_Powder.pdf
  - 12 ... https://www.auspigeonco.com.au/parasitic-diseases.html
  - 13 ... https://vetafarm.com.au/product/aviclens/
  - 14 ... https://vetafarm.com.au/product/avicare-conc-100ml/
  - 15 ... https://vetafarm.com.au/product/dnutrical-pdr-150g/
  - 16 ... https://vetafarm.com.au/product/multivet-liquid-50ml/
  - 17 ... https://www.nps.org.au/assets/medicines/9057bf75-9ebb-4773-8568-a53300feaebf.pdf
  - 18 ... https://betadine.com.au/product/betadine-antiseptic-topical-solution/
  - 19 ... https://morningbirdproducts.com/products/morning-bird-blood-stop-powder-natural-hemostatic-product
  - 20 ... https://www.wombaroo.com.au/product/good-oil-for-birds/
  - 21 ... https://nwlivestock.com.au/product/inca-sulfa-3/
  - 22 ... https://nwlivestock.com.au/product/ornithon-vitamin-supplement-for-birds/
  - 23 ... https://aristopet.com.au/products/oral-antibiotic-for-ornamental-birds/
  - 24 ... https://www.vetsense.com.au/product/vetsense-avicalcium/
  - 25 ... https://morningbirdproducts.com/products/antibacterial-antifungal-dietary-supplement
  - 26 ... https://vetafarm.com.au/product/triple-c/
  - 27 ... https://www.wombaroo.com.au/product/first-aid-for-birds/
  - 28 ... https://www.petindiaonline.com/story-details.php?ref=13956823
  - 29 ... https://www.aussiefinchforum.net/viewtopic.php?t=2716
  - 30 ... https://lafeber.com/pet-birds/questions/cleaning-cage/
  - 31 ... https://birdsupplies.com/blogs/news/bird-cage-cleaning-hacks
  - 32 ... https://www.mavlab.com.au/product/oxymav-b-for-birds-powder/
  - 33 ... https://petchemist.com.au/products/oxymav-b-soluble-antibiotic-powder-for-birds-100g.htmlx
  - 34 ... https://www.birdhealth.com.au/nutritional-supplements
  - 35 ... https://www.ranvet.com.au/products/virkon-s-4/
  - 36 ... The Finch Keepers Recipe Book, James P, 2010
  - 37 ... https://www.wombaroo.com.au/product/multi-clens/
  - 38 ... https://www.naturallyforbirds.com.au/shop/protein-boost-supplement-for-all-species
  - 39 ... https://www.naturallyforbirds.com.au/shop/micro-nutrients
  - 40 ... https://tummyrite.com.au/products/prosperity
  - 41 ... https://tummyrite.com.au/products/trp_fs
  - 42 ... https://www.vetsense.com.au/product/vetsense-birdcare-calcium-liquid/
  - 43 ... https://morningbirdproducts.com/products/morning-bird-calcium-plus-liquid-calcium-formula
  - 44 ... https://vetafarm.com.au/product/calcivet-50ml/
  - 45 ... https://www.wombaroo.com.au/product/liquid-gold/
  - 46 ... https://www.wombaroo.com.au/product/multi-vite-for-birds/
  - 47 ... Vetafarm - Data Sheet_Soluvite D.pdf
  - 48 ... Vetafarm - Data Sheet_Soluvite D Breeder.pdf
  - 49 ... https://www.naturallyforbirds.com.au/shop/prima-supplement-for-finches

INDEX

This is a comprehensive index, so it is suggested that this should be the first place to begin a "search" for information in this book.

Index

A

Abbreviations 137, 199, 216
Abnormal Head Movements 172
Activity Level 21
ACV. *See* Apple Cider Vinegar
Aggressive Birds 172, 184
Air Sac Mites 76, 85, 168
Air Sacs 191
Alfalfa Sprouts 235
Allele 191
Altricial 191
Amino Acids 56
Amino Acids and Vitamins 56
Animal Litter Floor 38
Annual Cleaning 74
Annual Feeding Timetable 67
Annual Plan 188, 205
Annual Timetable 96
Antibiotic 230
Anti-Biotic 85
Anti-Fungal 85
Antioxidants 91, 191
Appendices 225
Apple Cider Vinegar 91, 191, 229
Aspergillosis 75, 191
Assortative Mating 115, 191
Austerity Seed 52, 226
Australian Recessive Dilute 127, 133, 191, 200
Australian Recessive Dilute Blue 130
Australian Variegated Blue 130, 133, 145, 161, 191, 200
Australian White Breasted Yellow. *See* Australian Yellow
Australian Yellow 105, 124, 126, 133, 135, 141, 159, 191, 200, 223
Author 14
Autosomal 124, 135, 137
Autosomes 191, 216
AVG. *See* Australian Variegated Blue
Avian Paramyxovirus 172
Aviary 31, 176, 191
AviCalcium 229
Avicare 229
Aviclens 65, 229
Avi-Clot 229
Aviculture 192
Avicycline C 75
AY. *See* Australian Yellow

B

Back 17, 18
Back Colours 124
Bacterial Infections 75, 86
Baked Eggshells 54
Basic Diet 51
Bathing Facilities 25
Baycox 77, 90, 229
Bay Leaves 175
Beak 18, 192
   Colours 137
Behaviour 19
Behavioural Concerns 172
Belly 17, 18, 192
Benchmarks 202
Bengalese Finches 23, 110
Betadine 229
Bib 18, 192
Bicheno Finches 23

Bird Bath 43
BirdCare 57
Birdcare Calcium 227, 229
Bird Claw Trimmers 91, 192
Bird Dealers 118
Bird Expos 117
Bird Log 30, 120, 204
Bird Room Reference 217
Black-Headed 15, 124, 200
Blended Calcium 65
Blue-Backed 69, 103, 104, 125, 133, 135, 141, 160, 192, 200
   Health Problems 91
Boiled Eggs 64, 180, 226
Bok-choy 61, 226
Books 232
Breast Colours 17, 18, 124, 192
Breathing
   Clicked 192
Breeding 95, 200
   Cabinet 31, 42
   Floors 37
   Size 32
   Colonies 96
   Cycle 95
   Date Calculations 209
   Environment 97
   Issues 167
   Log 113, 210
   Maturity 17
   Next Cycle 108
   Pairs 96
   Period 183
   Plans 102, 103, 213
   Poor Results 174, 183
   Preparation 97

Records 112
Results 116
Season 192
Success 47
Timetable 108, 217
Tips 187
Breeding Plans 104
Broccoli 61, 91, 226
Broccoli Sprouts 235
Brood 192
Brooding 192
Bruno 192
Budgerigar Mix 52
Building Materials 39, 42, 232
Butcherbirds 169, 180

C

Cabinet 192. *See* Breeding Cabinet
Cage 31
  Hospital 27
  Transport 27
Cage Fright 117, 182, 192
Cage Front 192
Calcium 58, 90
Calcium Carbonate 54, 65
Calcium Concentration 226
Calcium & Iodine Bells 171
Calcium Plus 55, 227, 229
Calcium Supplements 53
Calcivet 55, 56, 79, 227, 229
Canary Seed 51, 226, 234
Candling Eggs 111
Carbohydrate Ratio 201
Carbohydrates 226
Care Plan 72
Carlox 77
Carotenoids 91, 163, 192
Carrots 61, 226
Catching Dividers 46
Catching Net 26
Chestnut Finches 23
Chicks 192, 195
Chicks in the Nest 109
Chicks per Season 115, 116

Chick Tossing 167, 184, 192
Chickweed 61, 226
Chicory 61
Chin 17, 18, 192
Chitted Seeds 66, 192
Chlorsig 229
Chlorsig Ointment 77
Choice of Mate 117
Chromosomes 18, 120, 134, 192
  Number 120
Cinnamon 127, 131, 133, 143, 192, 200
Circadian Rhythms 35
Claws 90, 192
Cleaning 36, 42, 73
  Annual 74
Clicked Breathing 192
Climate 28
Cloaca 18, 192
Clothing Moth Killers 176
Clutch 192
Clutch Ratio 201
Coccidiosis 77, 85, 87, 193
Cocciprol 77
Coccivet 77, 229
Cock 193
Cock or Hen 22
Cod Liver Oil 56
Co-dominant 125, 135, 193
Cold Draughts 33
Collar 193
Colouration 164
Colours 17
Combined Mutations 128, 136, 193
Compatibility 23, 183
Computer Records 114, 121
Concrete Floors 36
Coop Cups 44, 193
Coopex 76, 90, 98, 229
Cordon Bleu Waxbills 23
Corn 61, 91
Corona 193
Courtship 105
Crop 193

Crown 18, 193
Crows 180
Cryptoxanthin 91, 163, 165, 193
Cut-throat Finches 23
Cuttlebone 54, 193, 226
Cuttlebone Holder 25

D

Daily Essentials 56
Date Calculations 107, 209
Dead-in-Shell 193
Deaths of New Birds 182
Decision Tree for Health Diagnosis 206
Definitions 191
Design
  Aviary 176
  Breeding Cabinet 41, 42, 47
  Feeding Station 72
  Flight 39
  Hospital Cage 88
  Nest Box 98
  Nest Frame 99, 100
DF. *See* Double Factor
DF Pastel 193. *See* Yellow Back
Diamond Firetail Finches 23
Diarrhoea 193
Diatomaceous Earth 54, 76, 98, 175, 193
Diet 51
Dilute 126, 128, 133, 140, 193, 222
  Australian Recessive Dilute 127
  Euro Dominant Dilute 126
Dilute Blue 193
Dilute Gouldians 193
Dimensions 50
Dimorphic 193
Dirt Tray 193
Diseases 73, 86, 87
DNA Sexing 193
D Nutrical 60, 227, 229

D Nutrical Powder 55
Dominant 116, 124, 132, 193
Dosage Rates 83, 207, 229, 230, 231
Dosing
  Remote 83
Double-Bar Finches 23
Double-Factor 122, 134, 135, 193, 216
Double-Split 193
Doxycycline 75
Draught Shield 193
Droppings 193. *See also* Diarrhoea
Dry Floor 45
DufoPlus 55, 69, 229

E

Edstrom Water Valve 25, 43, 71, 194
Egg Binding 79, 87, 194
Egg & Biscuit 64, 226
Egg-Laying 107
Eggshells 54, 226
Eimeria Protozoa 77
Electrolyte for All Birds 57
Electrolytes 57
Electrolyte solutions 75
Endive 61, 226
Endocox 77
Enrofloxacin 75
Enteritis 75, 194
Environmental Stress 173
Epistatic Mutation 124, 194
E Powder Plus 229
Escaped Birds 185
Essential Care 187
Eumelanin 163, 194
Euro Dominant Dilute 126, 194
European Fallow 194
European Pastel 125, 194
Euro Yellow Back 125. *See* Yellow Back
Euro Yellow Backed 194
Excessive Heat 168

Excessive Melanin 78
Exhaust Fan 194
Eye Infections 77
Eyes 21

F

Facebook Group 189
Fail-Safe 43, 187, 194
Fallow 128, 131, 194
Fat 226
Fat Ratio 201
Feather Pigments 163
Feathers 194
  Condition 21
  Loss 86, 170, 182, 194
  Plucking 76
Feeding 51
  Stations 71
  Timetable 67
  Tips 69
Female Readiness 106
Fertility 111
Fibre Content 226
Fighting Pairs 184
Final Delivery Concentrations 60, 227
Final Vitamin Concentrations 227
Finch Club 189
Finch Mix 51, 194, 226
First Aid for Birds (Passwell) 229
Fit Grit 57, 65
Fledged Rate 201
Fledglings 108, 194
Flight 31, 194
  Size 32
Flocks 16, 180
Floors 36, 179
Floor Trays 44
Flow Chart for Treatment Decisions 82
Fluffed Up 194
Food Dispensers 24, 43
Foraging Trays 44, 63, 194
Foreign Body 173
Forms Library

Annual Plan for Gouldian Finches 205
Bird Log 204
Bird Room Reference 217
Breeding Date Calculations 209
Breeding Log 210
Breeding Plan 213
Decision Tree for Health Diagnosis 206
Dosage Rates 207
Health Treatment Notes 208
Offspring Log 214
Predicted Progeny 213
Punnett Squares 211, 212
Sale Transfer Form 118
Foster Parents 110, 117, 188
Fractures 77
French White Millet 51, 52, 226, 234
Full-Spectrum Lighting 35
Fungal Infections 75, 86, 173
Fungistat 76
F-Vite Plus 229

G

Gallery of Gouldian Colours 220
Garlic Powder 175
Gender Mixes 115
Genes 194
Genetic Collapse 194
Genetic Inheritance 134
Genetics 119, 200
  Australian Yellow 141
  Blue Backed 141
  Cinnamon 144
  Dilute 140
  Normal Gouldians 139
  Pastel Blue 142
  Seagreen 144
  Silver 142
  White-Breasted 140
  Yellow Back 140
Genetics Reference Table

139, 140, 141, 142, 143, 144, 145, 216
Genotype 120, 194
Genotype Conventions 123
Genotype Summaries 122
Getting Started 21, 27
Giardia 77, 87
Gizzard 194
Glossary of Terms 191
Going Light 80, 194
Goitre 171
Good Oil 231
Gouldian Nest Box 24
Gravel Floors 37
Green Food 194
Greens & Grains 52, 226
Grow Lights 170, 194

H

Habitat 16
Haldane's Rule 194
Hand-Feeding 194
Harkanker 77
Hatched Rate 201
Head 17, 100
Head Colours 115, 123, 132, 133
  Proportions 15
Head Movements 172, 183
Head Twirling 194. *See* Twirling
Health 73, 168, 200
Health Diagnosis 206
Health Issues 74
Health Treatment Notes 208
Healthy Birds 21
Heating 34, 40, 180
Hen 194
  Fertility 18
Heterologous 195
Heterozygous 120, 195
History of Gouldians 15
Homologous 195
Homozygous 120, 195. *See also* Double Factor
Hopper 70, 195. *See* Seed Hopper

Hospital Cage 27, 88
  Treatment Notes 89, 208
Hot Weather 35
Housing 31, 174, 187
  Alone 180
  Guideline 33, 179
Husk 186, 195
Hypercalcemia 58
Hyperthyroidism 58
Hypervitaminosis 57, 58
Hypocalcemia 58
Hypothyroidism 58
Hypovitaminosis 57, 58

I

Identifying Chicks 109
Illness
  Signs 21
Immune Function 195
Imprinting 111, 195
Inadequate Natural Light 78
Inbreeding 114, 173, 188, 195
Inbreeding Depression 114, 195
Incomplete Dominance 195. *See* Co-dominant
Incubation 108, 195
Independence 106, 108
Infections
  Bacterial 75, 86
  Fungal 75, 86
  Viral 75, 86
Inheritance
  Autosomal 135
  Back Colours 131
  Breast Colours 124
  Co-dominant Sex-Linked 135
  Head Colours 124
  Sex-Linked 134
Injuries 169, 183
Insects 181
Iodine 57, 58, 59, 90, 226, 230
  Recommended Concentration 59

Iodine Bells 171
Iodine Buttons 171
Iodine Concentration 226
Iodine Deficiency 170
Ioford 69, 229
Isobacterin 195
Isoxanthopterin 195
Ivermectin 76
Ivory 130, 145, 200

J

Japanese Millet 51, 52, 226, 234
Java Sparrows 23
John Gould 15
Juvenile 195, 224

K

Kale 61, 226
KD Powder 230
Keel Bone 195
Kimberley region 16

L

Lady Gould Finch 15, 195
Lavender Waxbills 23
Lebanese cucumber 61, 226
LED lights 35, 42
Leg Rings 28, 195
  Applicator 29
  Size 29
  Split or Fixed 29
Length of Gouldians 18
Lettuce 61, 226
Life Expectancy 18
Lighting 35, 42, 46, 117
Lilac Breasted 165
Lilac-Breasted 124
Lime 195
Line Breeding 114, 117, 195
Liquid Gold 55, 56, 227, 230
Liquid Iodine 171
Live Food 61, 177, 181
Liver Damage 69
Location & Aspect 34

Locus 195
Log
  Offspring 214
Long-Tail Finches 23
Lutein 91, 163, 165, 195
Lutino 124, 128, 131, 144, 195, 200

M

Maintenance 42, 47, 73
Male Gender Ratio 201
Male or female 22
Male Performance 105
Masked Finches 23
Mating Preferences 115
Mealworms 177, 226
Meanings 191
Measurement Conversions 94
Measurements 50
  Breeding Cabinet 41
  Perches 48
Medications 83
  Administration 83
  International Products 85
  Off-Market 84
  Safety Precautions 85
Medistatin 76
Melanin 195
Melanins 163
Melanism 78, 87
Melanocytes 195
Melba Finches 23
Mendelian Inheritance 119, 196
Mice 176
Micro-Nutrients 230
Millet sprays 63
Minerals 226
Miner Birds 169, 180
Mite Infestations 76, 86, 173
Mites 195
Mixed Collections 23
Mixed Seed 51
Monitoring 185
Monomorphic 196

Morph 196
Moth Balls 175
Moths in Seed Hoppers 174
Moulting 196
Moxidectin 76, 90, 230
Moxivet Plus 76, 230
Multi-Clens 66, 230
Multivet 56, 228, 230
Multivitamin 90
Multivitamins 55
Multi-Vite 55, 60, 227, 230
Mung Bean Sprouts 235
Mutation 104, 123
  Australian Recessive Dilute 127, 133
  Australian Variegated Blue 130, 145, 161
  Australian Yellow 105, 124, 126, 133, 135, 159, 165, 191
  Blue 129, 135, 165
  Blue-Backed 69, 103, 125, 133, 135, 160
  Breeding Plans 104
  Breeding Results 116
  Cinnamon 127, 131, 133, 166
  Colours 163
  Combinations 128, 193
  Dilute 133, 165
  Epistatic 124, 194
  Fallow 131
  Gallery 220
  Head Colours 132, 133
  Health Issues 81
  Ivory 130, 145
  Lutino 124, 128, 144
  Pastel Blue 106, 129, 131, 133
  Pigment Changes 165
  Prediction Software 138
  Satine 130, 145
  Seagreen 128, 131, 133, 166
  Secondary 129, 197
  Silver 107, 129, 131, 133
  White Breasted 157, 165

Yellow Back 105, 125, 128, 133, 135, 158, 165

N

Nail Clippers 91
Nape 17, 18, 196
Natural Earth Floors 36
Natural Wild-Types 22
Nesting Areas 45
Nesting Boxes 24, 98, 187, 196
  Available 184
  Inspections 184
  Installation 99
  Preparation 98
Nesting Material 26, 45
Nestlings 109, 196
Neurological Problems 173
Night Fright 173
Night-Fright 196
Nodules 109
Normal Gouldians 17, 139, 150, 196, 220
  Genetics 139
Northern Hemisphere 96, 205
Number of Birds 32
Nutmeg Mannikins 23
Nutrition 51, 200
Nutritional Deficiencies 78, 87
Nutritional Deficiency 172
Nutrition Facts 226
Nutrobal 55
Nystatin 76

O

Offspring Gender Ratio 201
Offspring Per Pair 202
Offspring Survival Rate 202
Omega 3 & 6 57
Online Discussion Group 189
Online Marketplaces 118
Open-Mouth Breathing 196
Orange-Headed 15

Ornithon 55, 60, 228, 230
Outcrossing 114, 196
Overcrowding 33
Over-Dosing 57
Owl Finches 23
Oxymav B 230

P

Painted Firetail Finches 23
Pair Bonding 117, 174, 196
Pairing 99
  Ages 100
  Close Relatives 99
  Examples 104
  Head Colours 100
  Non-recommended 103
Panicum Seed 51, 226
Paper Lined Floor 37
Paper Records 114, 121
Papillae 196
Paramyxovirus 196
Pastel 196
Pastel Blue 106, 125, 129, 131, 133, 142, 196, 200, 223
Pastel Body 129, 196. See Dilute
Pastel Green 126, 196
Pau D'Arco Bark Powder 76
Peas 61
Pecking Order 196
Perches 24, 48, 196
  Positioning 48
  Stress 197
Pet Stores 118
Phenotype 120, 196
Phenotype Summaries 121
Pheomelanin 163, 196
Pigment Changes 165
Pigments 163
Pine Shavings Floor 38
Plain Canary Seed 52
Plumage 196
Plumhead Finches 23
PMV. See Paramyxovirus
Polygenic Factors 196
Polymorphic 196

Poor Breeding Results 174
Popular Medications & Supplements 229
Predator-Aware 196
Predators 169, 173, 180
Predictable Progeny 101, 102, 211, 212, 213
Prediction Software 138
Preening 196
Preference Trials 234
  Sprouted Greens 235
Preventive Care 89
Prima 230
Products and Suppliers 232
Progeny 101, 138, 196
Progeny Predictions 150
Prosperity 230
Protection 31, 35
Protein 202, 226
  Nutrition 226
Protein Boost 56, 230
Protein Ratio 202
Pteridines 164, 197
Pumpkin 61, 91
Punnett Squares 146, 197, 211, 212
Purine 197
Purple Breast 124
Pyrethrin Spray 175

Q

Quarantining 23, 197
  Period 23
Questions & Answers 179
Quick of the Claw 90, 197
Quick Reference Table
  Annual Feeding Timetable 67, 217
  Diseases 86, 87
  Genetics of Gouldian Finches 139, 140, 141, 142, 143, 144, 145
  Gouldian Finch Progeny 150
  Head Colours in Gouldian Mutations 133

Recommended Measurements 50
Single & Double Factors in Mutations 135
Quik Gel 57, 75, 230

R

Rainbow Finches 15, 197
Ratio Benchmarks 202
Ratios 201
  Clutch Ratio 201
  Fledged Rate 201
  Hatched Rate 201
  Male Gender Ratio 201
  Offspring Gender Ratio 201
  Offspring Per Pair 202
  Offspring Survival Rate 202
  Protein Ratio 202
  Space Ratio 202
  Temperature Humidity Ratio 202
Recessive 124, 132, 197
Recommended Levels for Vitamins 57
Recommended Measurements 50
Recommended Reading 232
Recommended Resources 232
Record Keeping 28
Red-Brow Finches 23
Red-Faced Parrotfinch 23
Red-Headed 15, 124, 200
Red Panicum 52, 234
References 236
Remote Dosing 83
Remote Monitoring 185
Respiratory Infections 74, 86
Revive 76, 230
Rings. See Leg Rings
Rodents 176
Ronidazole 77
Ronivet-S 77
Roofing 169
Roosting 197

Routine Preventions 90
Rump 197

S

S76 76, 230
Saffron Finches 23
Sale Transfer Form 215
Sample Breeding Plans 104
Sand Floors 36, 37
Satine 130, 145, 197, 200
Satinet 197
Scatt 76
Seagreen 128, 131, 133, 144, 197, 200
Secondary Mutations 129, 136, 197, 216
Seed
   Austerity 52
   Greens & Grains 52
   Mixed 51
   Nutrition 226
   Seeding Grasses 63
Seed Hoppers 24, 43, 70, 174, 197
Seeding Grasses 63
Seed Preference 234
Seed Storage 53
Selling Offspring 117
Sex Chromosomes 120, 216
Sexing 22, 179, 197
Sex-Linked 124, 197
Sex-Linked Inheritance 134
SF Pastel 197. *See* Dilute
Shell Grit 25, 56, 197, 226
Silver 107, 129, 131, 133, 142, 197, 200, 224
Single-Factor 122, 134, 197, 200, 216
Sisal 26
Size 18
Snowpea Sprouts 61, 226, 235
Soaked Seed 65, 181, 226
   Procedure 65
Soaked Seeds 197
Society Finches 23
Soft Shell Eggs 183

Solaminovit 55, 56, 69, 228, 230
Soluvite D 55, 60, 171, 228, 231
Soluvite D Breeder 228
Solvita AD3 High E 69, 231
Southern Hemisphere 96, 205
Space per Bird 33, 41
Space Ratio 202
Spark 57, 75, 90, 231
Special Diets 69
Spice Finches 23
Spinach 61, 226
Spot-on Treatment 76, 83
Sprouted Greens
   Preference Trial 235
Sprouted Seed 66, 226
   Procedure 67
Sprouted Seeds 197
Star Finches 23
Star-Gazing 173, 197
Sternostoma tracheacolum 76, 168, 197
Stress Perches 28, 197
Structural Colours 163
Styptic Powder 91, 197
Subordinate Birds 197
Sulfa 3 75, 231
Sunlight 35, 78, 169, 176
Supplements 53
   Amino Acids 56
   Amino Acids and Vitamins 56
   Blue Gouldians 91
   Care Plan 72
   Final Concentrations 227
   Omega 3 & 6 57
   Vitamins 55
Suppliers 232
Survival of the Fittest 185
Survival Rates 115
Symbols 199

T

Tail 17, 198
Temperature Control 46

Temperature Humidity Ratio 202
Temperatures 28, 180
Terminology 19, 191
Territorial Disputes 19, 172
Territoriality 198
Testosterone Levels 19
Thank You 247
The Good Oil 56, 57, 60, 228
Thermostat 198
Thrush Infections 75
Thyroid 170
Timetable for Hatching 106
Tip of Beak 137
Torticollis 198. *See also* Twirling; *See also* Star-Gazing
Trait 198
Trait Abbreviations 216
Transfluthrin 176
Transport Cage 27
Trauma 77, 173
Tray 198
   Dirt 44, 193
   Foraging 44, 63, 194
   Wire 44, 198
Treatment Notes 89, 208
Trimming of Claws 90
Triple C 75, 231
TummyRite Plus 231
Turbobooster 56, 231
Twirling 172, 198

U

Under-Dosing 57
Undersized Chick 184
Useful Standard Dimensions 50
UV Exposure 78
UV Light 35, 169, 176

V

Vegetables 61, 181, 226
Vent 18, 198
Ventilation 35, 40, 42
Vermin Wire 40, 198

Index

Veterinarian 92
Video Monitoring 186
Viral Infections 75, 86
Virkon S 66, 231
Vitamin A 58, 91, 172
Vitamin B Complex 58
Vitamin D3 55, 58, 169, 176
Vitamin D Deficiency 78, 87, 169
Vitamin D Supplements 170
Vitamin D Synthesis 35
Vitamin E 58
Vitamin K 58
Vitamins 55
   Final Concentrations 227
   Other Sources 59
   Overdosing 69
   Recommended Concentrations 57
   Undisclosed Concentrations 60
Vitamin Supplements Recommended 57
Vitiligo 79, 198
Volume Conversions 94

W

Watercress 61
Water Dispensers 25, 43, 70
Automatic 71
Bottles 71
Bowls 70
Fountains 71
Valves 71
Weather Protection 34
Webbing 198
Webbing Moths 174
Websites 232
Weight Conversions 94
Weight of Gouldians 18
Wellness 73
Wet Season 96
White-Breasted 124, 140, 157, 200
   Blue 129
   Genetics 140
   Silver 130
   Yellow Back 128
Wild-Type 150, 198; *See also* Normal Gouldians
Wimple 198. *See* Nape
Wind Protection 33
Wing 198
Winnowing 186, 198
Wire. *See* Vermin Wire
Wire Mesh Floors 37, 38
Wire Tray 198
Wooden Floors 36

X

Xanthopterin 198

Y

YB. *See* Yellow Back
Yellow Back 105, 125, 128, 133, 135, 158, 198, 200, 221
Yellow-Backed 185
   Genetics 140
   Identification 127
Yellow-Headed 15, 124, 200

Z

Zade 56, 69, 231
Zeaxanthin 91, 163, 165, 198
Zebra Finches 23
ZolCal D 79
ZolCal-F 56

A Special Thank You

The author has benefited from the advice of a large number of different people over his 50 year experience with Gouldian Finches. With too many to list them all by name, their generous sharing of knowledge is sincerely appreciated.

Gouldian Finches - Care, Breeding & Genetics

© 2025 - Hanks, Tony

Published by Gouldian Care

www.gouldiancare.com

www.ingramcontent.com/pod-product-compliance
Lightning Source LLC
Chambersburg PA
CBHW042229090526
44587CB00001B/1